Persistent Organic Pollutants (POPs) - Monitoring, Impact and Treatment

Edited by Mohamed Nageeb Rashed

Published in London, United Kingdom

IntechOpen

Supporting open minds since 2005

Persistent Organic Pollutants (POPs) - Monitoring, Impact and Treatment
http://dx.doi.org/10.5772/intechopen.95151
Edited by Mohamed Nageeb Rashed

Contributors
Giovanni Cagnetta, Mohammadtaghi Vakili, Siyabonga Aubrey Mhlongo, Kholofelo Clifford Malematja, Linda Lunga Sibali, Peter P. Ndibewu, Ifeoluwa Oluwafunmilayo Daramola, Matthew Ayorinde Adebayo, Amany El-Sikaily, Mohamed Helal, Sara Ghanem, Nisha Gaur, Dhiraj Dutta, Aman Jaiswal, Rama Dubey, Dev Vrat Kamboj, Yesim Gucbilmez, Ahmed Nasri, Patricia Aïssa, Ezzeddine Mahmoudi, Hamouda Beyrem, Véronique Perrier, Poslet Morgan Shumbula, Collet Maswanganyi, Ndivhuwo Shumbula, Takoua Mhadhbi, Mohamed Allouche, Amel Hannachi, Hary Demey

First published in London, United Kingdom, 2022 by IntechOpen
IntechOpen is the global imprint of INTECHOPEN LIMITED, registered in England and Wales, registration number: 11086078, 5 Princes Gate Court, London, SW7 2QJ, United Kingdom
Printed in Croatia

British Library Cataloguing-in-Publication Data
A catalogue record for this book is available from the British Library

Additional hard and PDF copies can be obtained from orders@intechopen.com

Persistent Organic Pollutants (POPs) - Monitoring, Impact and Treatment
Edited by Mohamed Nageeb Rashed
p. cm.

This title is part of the Environmental Sciences Book Series, Volume 1
Topic: Pollution
Series Editor: J. Kevin Summers
Topic Editors: Ismail Md. Mofizur Rahman and Zinnat Ara Begum

Print ISBN 978-1-83968-206-3
Online ISBN 978-1-83968-207-0
eBook (PDF) ISBN 978-1-83968-208-7
ISSN: 2754-6713

For EU product safety concerns:
IN TECH d.o.o., Prolaz Marije Krucifikse Kozulić 3, 51000 Rijeka, Croatia, info@intechopen.com or visit our website at intechopen.com.

IntechOpen Book Series

Environmental Sciences

Volume 1

Aims and Scope of the Series

Scientists have long researched to understand the environment and man's place in it. The search for this knowledge grows in importance as rapid increases in population and economic development intensify humans' stresses on ecosystems. Fortunately, rapid increases in multiple scientific areas are advancing our understanding of environmental sciences. Breakthroughs in computing, molecular biology, ecology, and sustainability science are enhancing our ability to utilize environmental sciences to address real-world problems.

The four topics of this book series - Pollution; Environmental Resilience and Management; Ecosystems and Biodiversity; and Water Science - will address important areas of advancement in the environmental sciences. They will represent an excellent initial grouping of published works on these critical topics.

Meet the Series Editor

J. Kevin Summers is a Senior Research Ecologist at the Environmental Protection Agency's (EPA) Gulf Ecosystem Measurement and Modeling Division. He is currently working with colleagues in the Sustainable and Healthy Communities Program to develop an index of community resilience to natural hazards, an index of human well-being that can be linked to changes in the ecosystem, social and economic services, and a community sustainability tool for communities with populations under 40,000. He leads research efforts for indicator and indices development. Dr. Summers is a systems ecologist and began his career at the EPA in 1989 and has worked in various programs and capacities. This includes leading the National Coastal Assessment in collaboration with the Office of Water which culminated in the award-winning National Coastal Condition Report series (four volumes between 2001 and 2012), and which integrates water quality, sediment quality, habitat, and biological data to assess the ecosystem condition of the United States estuaries. He was acting National Program Director for Ecology for the EPA between 2004 and 2006. He has authored approximately 150 peer-reviewed journal articles, book chapters, and reports and has received many awards for technical accomplishments from the EPA and from outside of the agency. Dr. Summers holds a BA in Zoology and Psychology, an MA in Ecology, and Ph.D. in Systems Ecology/Biology.

Meet the Volume Editor

Prof. Mohamed Nageeb Rashed is a Professor of Analytical and Environmental Chemistry and former vice-dean for environmental affairs, Faculty of Science, Aswan University, Egypt. He received his Ph.D. in Environmental Analytical Chemistry from Assiut University, Egypt, in 1989. His research interest is in analytical and environmental chemistry with special emphasis on: (1) monitoring and assessing biological trace elements and toxic metals in human blood, urine, water, crops, vegetables, and medicinal plants; (2) relationships between environmental heavy metals and human diseases; (3) uses of biological indicators for monitoring water pollution; (4) environmental chemistry of lakes, rivers, and well water; (5) water and wastewater treatment by adsorption and photocatalysis techniques; (6) soil and water pollution monitoring, control, and treatment; and (7) advanced oxidation treatment. Prof. Rashed has supervised several MSc and Ph.D. theses in the field of analytical and environmental chemistry. He served as an examiner for several Ph.D. theses in analytical chemistry in India, Kazakhstan, and Botswana. He has published about ninety scientific papers in peer-reviewed international journals and several papers in national and international conferences. He participated as an invited speaker at thirty international conferences. Prof. Rashed is the editor-in-chief and an editorial board member for several international journals in the fields of chemistry and environment. He is a member of several national and international societies. He received the Egyptian State Award for Environmental Research in 2001 and the Aswan University Merit Award for Basic Science in 2020. Prof. Rashed was recognized in Stanford University's list of the World's Top 2% Scientists in 2020 and 2021.

Contents

Preface

Persistent organic pollutants (POPs) are toxic organic compounds that are resistant to environmental degradation through biological, chemical, and photolytic processes. The most used POPs are polychlorinated biphenyls (PCBs), polycyclic aromatic hydrocarbons (PAHs), dichlorodiphenyltrichloroethane (DDT), lindane, dieldrin, and dioxins. Many POPs are currently used as pesticides, pharmaceuticals, solvents, and industrial chemicals. Because of their persistence, POPs bioaccumulate and affect human health and the environment worldwide. They can be transported by wind and water from one country to another and thus can affect people and wildlife far from where they are used and released. POPs persist for long periods of time in the environment and can accumulate and pass from one species to the next through the food chain, accumulating in the body fat of living organisms.

This book gathers recent research by outstanding experts in monitoring and treatment technologies of POPs as well as their effects on human health and the environment. *Persistent Organic Pollutants (POPs) - Monitoring, Impact and Treatment* deals with several aspects of POP monitoring, occurrence, impact, and treatment technologies. The book is divided into two sections containing nine chapters. The first section covers monitoring, impacts, and occurrence in four chapters: Chapter 1 "Endocrine Disruptor Impact on Zebrafish Larvae: Posterior Lateral Line System as a New Target", Chapter 2 "Role and Effect of Persistent Organic Pollutants to Our Environment and Wildlife", Chapter 3 "Impact of PCBs, Furan and Dioxin on Hepatocarcinogenesis", and Chapter 4 "Persistent Organic Pollutants in the Bizerte Lagoon Ecosystem: Occurrence, Distribution, and Ecotoxicological Assessment Using Marine Organism". The second section explains the different techniques used for the treatment of POPs in five chapters: Chapter 5 "Type, Sources, Methods and Treatment of Persistent Organic Pollutants in Wastewater", Chapter 6 "Nonthermal Mechanochemical Destruction of POPs", Chapter 7 "Physiochemical Properties and Removal Methods of Phenolic Compounds from Waste Waters", Chapter 8 "Occurrence and Removal of Persistent Organic Pollutants (POPs)", and Chapter 9 "Recent Developments in the Application of Advanced Oxidative Processes for Remediation of Persistent Organic Pollutants from Water." This volume is a useful resource for researchers and professionals working with POPs.

I would like to express our sincere thanks to Dr. Sally Salaah Eldin Elshalqamy, National Institute of Oceanography and Fisheries (NIOF), Alexandria Egypt, and Dr. Sawsan Elateek, Faculty of Agriculture, Ain Shams University, Egypt, for their contribution to the scientific revision of the book's chapters. I would also like to express my thanks to the contributing authors for their hard work and Author Service Manager Ms. Marijana Josipovic at IntechOpen.

Dr. Mohamed Nageeb Rashed
Professor,
Faculty of Science,
Aswan University,
Aswan, Egypt

Section 1

Occurrence and Monitoring of POPs

Chapter 1

Endocrine Disruptor Impact on Zebrafish Larvae: Posterior Lateral Line System as a New Target

Ahmed Nasri, Patricia Aïssa, Ezzeddine Mahmoudi, Hamouda Beyrem and Véronique Perrier

Abstract

Endocrine-disrupting chemicals (EDCs), including polychlorinated biphenyls (PCBs), bisphenol A (BPA), pharmaceutical drugs, and pesticides, affect a variety of hormone-regulated physiological pathways in humans and wildlife. The occurrence of these EDCs in the aquatic environment is linked with vertebrates' health alteration. EDCs exhibit lipophilic characteristics and bind to hydrophobic areas of steroid receptors, such as the estrogen receptor, which are involved in vertebrate developmental regulation. Mainly, EDCs modify the transcription of several genes involved in individual homeostasis. Zebrafish conserve many developmental pathways found in humans, which makes it an appreciated model system for EDCs research studies, especially on early organ development. In the current chapter, we emphasize on latest published papers of EDCs effects on lateral line regeneration in zebrafish larvae. Similarly, we describe other special impacts of EDCs exposure. In conclusion, we make the case that the zebrafish lateral line exposed to EDCs can provide important insights into human health.

Keywords: zebrafish larvae, posterior lateral line (PLL), endocrine-disrupting chemicals (EDCs), experimental exposure, regeneration

1. Introduction

During the last two decades, special attention from scientists and the public has been increasingly given to the harmful effects that may result from exposure of humans or wildlife to chemicals, having the property of interfering with the endocrine system. According to McLachlan John [1], one of the very first references to the problematic posed by what are called "endocrine disruptors" is that of Roy Hertz [2] who considered "that the fact should be taken into account that the use of hormones in animal feed risked exposing certain individuals to these substances, when they should never have been in contact with such molecules in their lifetime, that we were in the process of creating a steroid cycle in our environment, and that we had to seriously consider the implications this could have for our development, our growth and perhaps for our reproductive function."

The presence of endocrine disruptors (EDCs) in the environment has raised many questions within the scientific community because of the risks they represent for humans and ecosystems [3]. Indeed, EDCs can cause "adverse effects on the health of an organism or its progeny in relation to changes in endocrine function" [4]. EDCs can act on all stages of endocrine regulation, from the hormones synthesis to their action in target tissues [3]. They include a large category of chemical substances such as natural and synthetic hormones or synthetic molecules such as pesticides, polychlorinated biphenyls (PCBs), or even alkylphenols from the industrial activities increasing as well as the pharmaceuticals used in everyday life.

Aquatic environments are often the ultimate outlet for many anthropogenic chemicals [5]. Several studies, carried out in the laboratory or in the natural environment, have revealed harmful effects of these substances on fauna and in particular on the physiology of fish reproduction [6]. Numerous studies have established a direct link between the presence of EDCs in the environment and alterations in reproductive function in certain aquatic organisms [6]. In fish, disturbances of hormone-regulated proteins, as well as histological alterations in the male gonads, have been demonstrated and linked to exposure to estrogen mimetics [7]. Indeed, certain EDCs are compounds that act as agonists of estrogen receptors (ERs) such as 17α-ethinylestradiol (EE2). These estrogen-mimetic compounds also include certain pesticides, alkylphenols, or bisphenols. These EDCs can thus mimic the action of endogenous natural estradiol and are able to induce the expression of estrogen-regulated genes [8].

While most of the research has focused on the EDCs effects on peripheral organs and more specifically on the gonad [9], this chapter aims to focus on the latest published research on the impact of EDCs on the peripheral nervous system of zebrafish larvae, more precisely, the posterior lateral line PLLs regeneration after exposure to environmental pollutants.

2. Posterior lateral line (PLL)

2.1 Definition and organization

The mechanosensory lateral line system is found in more than 25,000 fishes species—the Chondrichthyes (sharks, skates, rays, and chimeras), the "Agnatha" (the jawless fishes, the hagfishes, and lampreys), and the Osteichthyes (the bony fishes). The lateral line systems of the Osteichthyes and Chondrichthyes are more evolved, with superficial neuromasts on the skin and canal neuromasts in a series of lateral line canals on the head and trunk. The lateral line system of the lampreys is composed exclusively of superficial neuromasts located in grooves on the skin of the head and trunk. The lateral line system is a mechanosensory organ that detects the water vibration up to about 300 Hz [10], and relative pressure to the body. The information collected will be used for the detection of prey, schooling behavior, and avoiding predators [11]. Thus, the lateral line is composed of mechanoreceptors organs called neuromasts, which are distributed over the fish body. Neuromasts comprise a core of sensory hair cells surrounded by nonsensory support cells (**Figure 1**). Each neuromast is innervated by a branch of the lateral line nerve (axonal nerve). Generally, the lateral line is subdivided in two—the anterior lateral line (ALL) which comprises the neuromasts present on the head, jaw, and opercle, and its sensory neurons form the ALL ganglion, rostral to the ear, and the posterior lateral line (PLL) which contains the neuromasts on the trunk and tail and its sensory neurons form the PLL ganglion, caudal to the ear. Both the ALL and the PLL comprise several branches.

Each lateral line neuromast contains a central sensory hair cell progenitor embedded within a rosette formed by apical attachment and constriction of surrounding epithelial support cells. Lateral line hair cells are surrounded by a

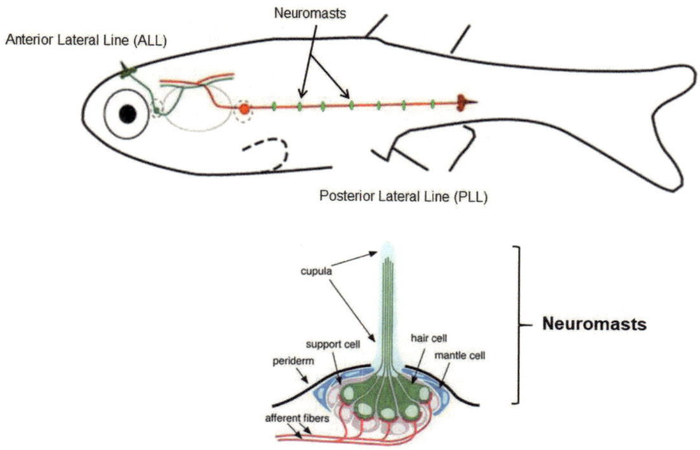

Figure 1.
Zebrafish posterior lateral line organization (PLL).

group of support cells. Many of these cells are located basally to the hair cells and project interdigitating processes between them, acting to isolate hair cells from one another. In addition to serving as the source for new hair cells in the event of damage, they also provide structural and trophic support. Lateral line hair cells also share structural, functional, and molecular similarities with the hair cells in the vertebrate inner ear [12]. They are innervated by both afferent and efferent nerve fibers on their basal surfaces, which emanate from the lateral line ganglion and branch off at each neuromast [13]. The lateral line system has become increasingly popular as a model for studying hair cell biology relating to human hearing and balance disorders (**Figure 1**).

2.2 PLL development

In zebrafish, the posterior lateral line system genesis (PLL) is started at the end of 24 hpf by the migration of cells cluster (100 cells) called the primordium PLL (PLLp) under the skin near the ear to the end of the tail [14]. During this step, the PLLp periodically deposits neuromasts (L1, L2 ... L7) along the body and will finish its migration, by the establishment of 2–3 terminal neuromasts (TN) at the level of the tail. In addition, each neuromast is formed by the sensory hair cells in the center. The pLLP is prearranged along its migratory axis—the posterior third cells (head zone) are highly proliferative and mesenchymal type, while those of the anterior two-thirds (leakage zone) are placed in epithelial rosettes [15]. These rosette cells called protoneuromasts will give rise to hair cells or support cells after they are deposited (**Figure 2**).

The sensory hair cells are formed from the central progenitor hair cells which after division produce pairs of differentiated hair cells sensitive to water movement. Support cells surround the central sensory hair cells and provide structural support [16]. Also, these internal support cells are themselves surrounded by mantle cells or external support cells. These cells form internal support cells during their proliferation (**Figure 1**) [17].

2.2.1 Genetic control

The PLLp migration is genetically regulated by several genes including those of the chemokine Cxcl12a and its receptors Cxcr4b and Cxcr7b. Cxcl12a is expressed in cells along the horizontal myoseptum, while PLLp expresses both Cxcr4b and

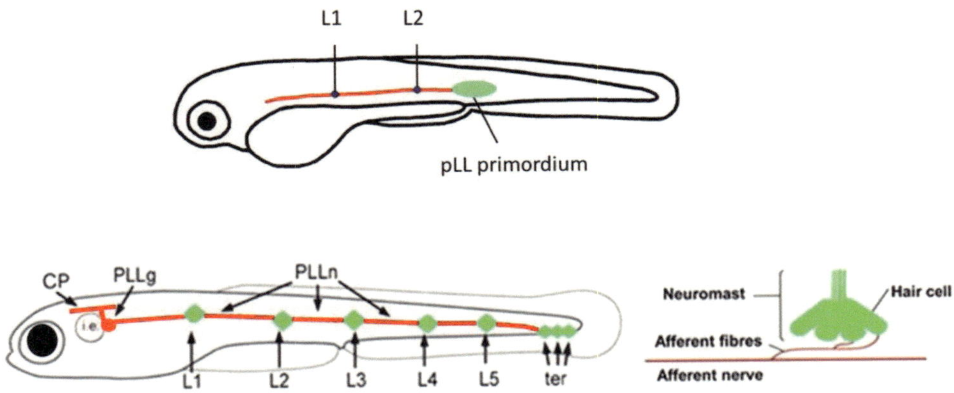

Figure 2.
Zebrafish neuromast anatomy and posterior lateral line development.

Cxcr7b. The PLLp migration is inhibited following disruption of expression of Cxcl12a and these receptors [18]. Dambly-Chaudière and Ghysen [19] have shown that from an affinity point of view, Cxcr7b has more affinity for Cxcl12a (expressed in the trailing zone) than for Cxcr4b (expressed in the head zone). Cxcr7b to serve as a molecular sink due to this difference in binding affinity, preventing Cxcr4b receptors expressed near the leak area from binding to the chemokine [20]. Thus, through the primordium an expression gradient of Cxcl12a is generated, the binding of which can involve the polymerization of actin in the direction of migration [21].

The PLLp migration is dependent on the canonical signaling interaction between Wnt and FGF. In the head zone, the Wnt signaling is important while in the tail zone it is the FGF signaling which is essential. Wnt signaling results in Cxcr4b expression and proliferation mediation in the head area, allowing the primordium to maintain its size throughout its migration [22]. These proliferating cells migrate from the head area and grow throughout the pPLL, thus depositing several protoneuromasts. The number of neuromasts generated is reduced and the speed of neuromast placement decreases following disruption of proliferation [23, 24]. Wnt signaling also controls the expression of Fgf3 and Fgf10a ligands [25]. FGF signaling drives the morphogenesis of epithelial rosettes which will give rise to neuromasts at the level of the leakage zone [26]. Studies have shown that inhibition of FGF signaling inhibits the formation of rosettes, and consequently the formation of neuromasts [27]. Recently, Yanicostas et al. [28] reported that primordial migration is inhibited following the inactivation of ka1la, which is a homologous zebrafish gene encoding the extracellular matrix protein Anosmin-1a and known to be an activator of FGF signaling [29]. Likewise, the expression of ka1la is similar to that of cxcr4b, important in the head region and less essential in the tail region, but independent of CXCR4b/SDF1a signaling.

2.2.2 Estrogen receptors implication

In zebrafish, the most studied nuclear receptors are estrogen receptors (ERs). Three genes encoding these receptors have been identified—one encoding an ERα ortholog in mammals and two orthologs encoding ERβ (called ERβ1 and ERβ2) [30]. These receptors are found scattered throughout several regions of the body such as the gonads, liver, and nervous system [30]. Several studies have shown that all estrogen receptor isoforms exhibit high expression levels specifically in lateral line neuromasts [31]. Research work on the importance of ERs in the establishment of PLL has shown that a disturbance in the development of neuromasts by the absence of hair cells, occurs following the temporary suppression of the expression of ERβ2 by a morpholino, which could be linked to aberrant activation of the Notch signaling pathway in embryos [32]. The temporary suppression by a morpholino of the expression of ERβ2 led to the disturbing development of these neuromasts (absence of hair cells), which could be related to aberrant activation of the Notch signaling pathway in embryos treated with morpholino [32]. Likewise, developmental defects and early embryo mortality occur following the suppression of ERα expression by application of morpholino from the translation of maternal transcripts [33]. Recently, a mutation in the gene encoding ERβ2 made it possible to identify a new mutant zebrafish line [34]. Deformed sexual intercourse (dominance of the adult male population), testes of altered morphologies, an imbalance in hormone levels, and an altered immune system, are the results of this mutation [34].

3. Axonal nerve PLL

3.1 Axonal nerve regeneration

Several tissues in fish exhibit a remarkable capacity for regeneration after injury, including the retina, cardiac tissue, and neurons. The lateral line which is a sensory system located on the surface and used to detect the movement of water shows a robust regenerative capacity. In addition, O'Brien et al. [35] approved that all cell types in this system can be genetically, physically, or chemically modified. Neuromasts deposited on the body surface of zebrafish larvae are innervated by sensory axons (PLL nerve) [11]. The superficial development of the nerve allows localized lesion, thus, the dynamics of complete regeneration of axons has been studied in zebrafish larvae 24 hours after axotomy [36]. Several works have studied the involvement of different cell types in the dynamics of degeneration and regeneration of the lateral line nerve (PLL). The inhibition of Schwann cell expression after chemical depletion inhibits the binding of neuregulin to the Erb receptor, which causes the exhaustion of nerves in peripheral Schwann cells [37]. In addition, the use of a transgenic mutant fish "named leo1" (no development of neural crest derivatives according to Nguyen et al. [38] leads to incoherent axonal regeneration. Also, inhibition of macrophages (by morpholino: molecule used to modify gene expression) causes very slow regeneration of the PLL nerve (more than 6 hours). In doubly mutant individuals characterized by the absence of Schwann cells and macrophages, axotomy of the nerve is followed by the death of these individuals [36].

The PLL nerve undergoes Wallerian degeneration (WD) after axotomy (**Figure 3**) with a succession of three phases; a delay phase (phase 1), a fragmentation phase (phase 2) followed by a final clearance phase (phase 3). The two phases of fragmentation and clearance begin approximately 3–5 hours post-axotomy respectively. In zebrafish, Wallerian degeneration (WD) of posterior lateral line axons occurs much faster than that studied in mammals and Drosophila [36]. Wallerian degeneration (WD) occurs in an orderly and stereotypical fashion involving described genetic control in the central nervous system (CNS) and peripheral (PNS) after trauma, stroke, or infection [39]. After axon cleavage, acute axonal degeneration (AAD) can occur at both adjacent ends of the lesion [40]. After ADA, it has been reported that the detached fragment remains intact during the delay phase or "lag phase." Following this latency phase, the axons will rapidly degenerate followed by a cut in the endoplasmic reticulum, degradation of the neurofilament, and swelling of the mitochondria of the axonal fragments. During the final phase, these fragments are removed using phagocytic cells. At the level of the PLL system, Schwann cells and macrophages play an important role in the process of Wallerian degeneration (WD). Thus, Schwann cells decrease

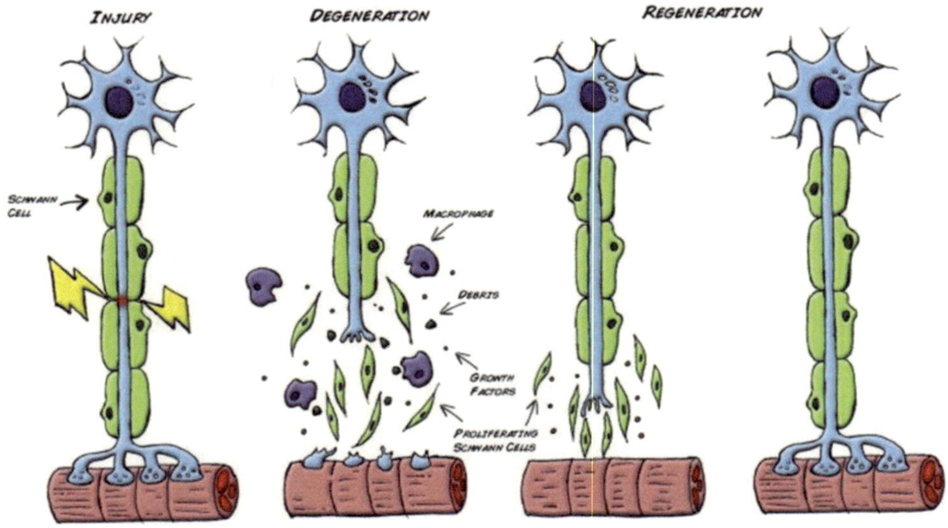

Figure 3.
Wallerian degeneration (WD) steps.

myelin lipid synthesis in the first 12 hours after axotomy [41] and inhibit myelin protein production for 48 hours [42].

4. Sensory hair cells

The hair cells of neuromasts are mechanosensory cells that are able to detect the water movements and transform the energy generated into electrical signals transmitted automatically to the brain. Usually, the hair cell is highly polarized, both apicobasal and in the plane of the epithelium. In its structure, it is characterized by a crescent-shaped stereociliary bundle and a large single kinocilium, on the apical side of the cell facing the otic lumen (**Figure 4**). Due to the morphology and function of hair cells in the lateral line system, these cells are very similar to those in the inner ear of mammals [12]. Numerous researchers have demonstrated that lateral line hair cells are sensitive to exposure to aminoglycosides [43, 44], *in vivo* imaging of fish lateral line hair cells zebra was first used by [45] who observed the death and regeneration of hair cells induced by neomycin. Harris et al. [17] then developed additional assays to quantify the death and regeneration of hair cells in the lateral line, establishing a basis for genetic and chemical studies aimed at detecting modulators of hair cell sensitivity to ototoxic exposures and to perform further testing. Other research works [46] have been studied the lateral line response of zebrafish following exposure to ototoxic compounds, such as aminoglycosides and cisplatin. The zebrafish lateral line system is, therefore, a rapid and efficient model for evaluating the effects of a large number of pharmaceuticals on mechanical-sensory hair cells [32].

4.1 Sensory hair cell regeneration

4.1.1 Cell differentiation and proliferation

Neuromast hair cells are functional in zebrafish 3 days after fertilization [19] and contain 8–20 hair cells 5 days after fertilization [17]. They are surrounded by

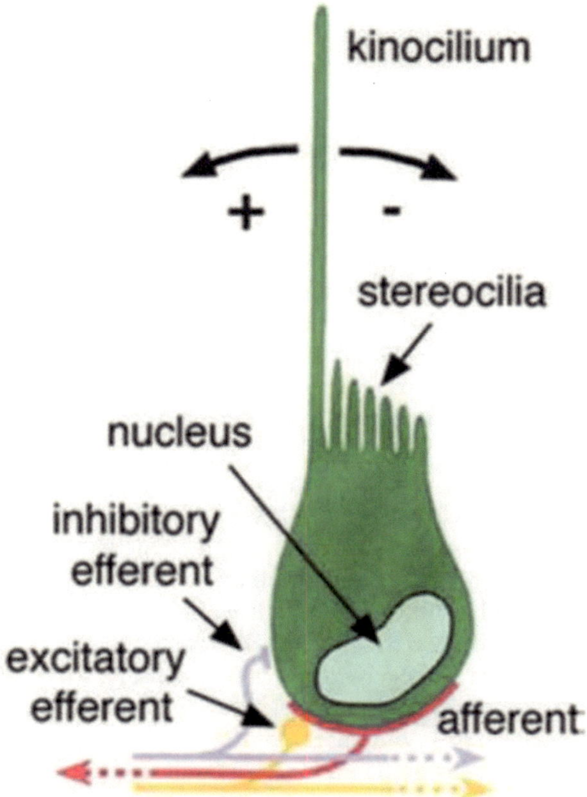

Figure 4.
Hair cell of neuromast showing its functional asymmetry and its afferent/efferent innervation.

nonsensory support cells, with basal nuclei and apical projections that intersect between them [47]. Rubel et al. [48] report that hair cells regenerate after damage via trans-differentiation or proliferation of carrier cells. Studies performed using tritium and bromodeoxyuridine (BrdU) labeling techniques have shown that lateral line hair cells can undergo continuous proliferation [49]. After acoustic trauma, fish can regenerate hair cells within 1–2 weeks [50]. However, the regenerative potential of neuromast hair cells can be considered a dose-dependent response depending on the level of damage [51]. Several studies have shown that hair cells can regenerate from mitotic divisions and the proliferation of supporting cells. At the zebrafish lateral line, hair cells normally undergo programmed cell death during development but are restored from support cells to the periphery after the S phase is produced [45]. The proliferating supporting cells can either remain on the periphery or migrate inward and their number increases after drug-induced hair cell death. Further studies on the zebrafish lateral line have shown that the newly formed hair cells are the result of the proliferation of supporting (carrier) cells and that there are two sets of these cells within the neuromasts [52]; one group of cells is centrally located and considered the progenitor of hair cells, the other is peripheral whose function is unknown. This suggests that there may be functional specializations between populations of neuromast support cells. Hair cells can also regenerate from the trans-differentiation of carrier cells. By applying high levels of damage to neuromasts, hair cell replacement allows surrounding support cells to divide [53].

4.1.2 Regeneration mechanisms

The determination of genes and molecular mechanisms controlling the regeneration of hair cells via differentiation and proliferation of support cells has received great interest. Recently, DNA microarrays and next-generation sequencing (high throughput sequencing) have been used to identify which genes are activated after the destruction of hair cells. Thus, following the exposure of zebrafish to noise, a DNA chip was produced in order to follow the change in gene transcripts [54]. A modification of the genes encoding growth hormone and genes for myosin (light and heavy chains) and the major histocompatibility complex have been observed. Liang et al. [55] have shown that the "stat3/socs3" pathway can modulate the production of lateral line hair cells during development and of the adult inner ear during regeneration. In mouse models, Stat3 effectors may be involved in hair cell survival [56], but no role of Stat3 in hair cell regeneration has yet been reported in mammals.

Chemical screening techniques (chemical screening) have also been used to identify compounds that increase or inhibit hair cell regeneration in the lateral line of zebrafish. Chemical screening has been used to identify synthetic glucocorticoid activators that promote hair cell regeneration by increasing mitotic activity [57]. This study also identified inhibitors that reduced hair cell regeneration or prevented cell proliferation. Research on zebrafish has shown that the "Wnt" signaling pathway is involved in the regeneration of hair cells. At the neuromast level, inhibition of "Wnt/β-catenin" signaling reduces proliferation and differentiation of hair cells while activation of "Wnt" increases the number of hair cells and promotes reintegration of support cells in the cycle of hair cells and their proliferation [58]. In addition, activation of Wnt/β-catenin causes increased regeneration of hair cells [59]. The size of the neuromast is also regulated by a negative feedback loop that integrates the "Wnt" signaling activity [60] and promotes the proliferation of surrounding cells.

Jiang et al. [61] showed that analysis of RNA transcripts expressed in the zebrafish lateral line following neomycin-induced damage showed that Wnt/β-catenin signaling is weakly regulated at onset, but becomes highly regulated later, suggesting that "Wnt" is necessary for hair cell proliferation, but not immediately after hair cell damage. The research focused on "Wnt" signaling in zebrafish indicates that the "Sox2" transcription factor is involved in the proliferation and trans-differentiation of hair cells [59]. Thus, it has been shown that the newly formed hair cells originate from the proliferation of "Sox2 positive" cells. The "Sox2" factor is highly expressed in most progenitor cells of proliferating neuromasts [53] and is also required for trans-differentiation of carrier cells [62]. Another regulator of "Wnt/β-catenin" signaling is "ErbB/Neuregulin," which may act to regulate zebrafish lateral line interneuromast cells proliferation and neuromast development [63]. Other work suggests that the "Notch" pathway modulates regeneration because inhibition of "Notch" signaling can cause regeneration of hair cells [64]. Jiang et al. [61] studying lateral line hair cell regeneration in zebrafish found that "Notch" signaling is inhibited immediately after hair cell damage. In addition, constitutive expression of "Notch" may prevent the proliferation of hair cells from carrier progenitor cells, while the elimination of "Notch" activity produces an increased number of cellular progenitors and hair cells [16]. Wada et al. [60] reported that the proliferation pathways of hair cells, Wnt, Notch, and Erb are key components common in zebrafish and mouse models involved in the regulation of hair cell regeneration. In addition, the cell cycle inhibitor p27kip is known to regulate the regeneration of mammalian hair cells [65].

4.2 Regeneration after contaminant exposure

The lateral line system has been used to study the various compounds' toxicity, including aminoglycosides [44], trace metals [51], cisplatin [66], and endocrine disruptors [67–69] during the zebrafish early life stages. The hair cells death by necrosis and apoptosis, the new hair cells differentiation and proliferation following exposure to toxic substances have been described by several studies [44, 51, 66–69].

4.2.1 Metal

Hernandez et al. [51] have shown that copper is toxic to the zebrafish lateral line following exposure to concentrations varying between 1 and 50 M resulting in the death of hair cells of neuromasts. During the first 5 minutes of copper sulfate "CuSO4" exposure (5 M), they appear signs of damage [70]. Thus, morphological changes resulting from the onset of apoptosis and necrosis have been recorded [51]. Exposure of zebrafish larvae to "CuSO4" for 2 hours, followed by assessment of hair cell regeneration over the next 5 days, resulted in robust regeneration in neuromasts of the anterior lateral line, whereas posterior lateral line neuromasts showed little regeneration, suggesting a differential regeneration in the lateral line within the same animal [51]. In other work, they have shown that copper attenuates hair cell regeneration in part by reducing cell proliferation [71]. Also, neuromasts did not regenerate upon continuous exposure to copper [71]. These results suggest that copper is toxic to both hair cells and support cells, and its presence in waterways can negatively affect fitness. Other work has shown that the exposure of zebrafish larvae to cadmium causes an alteration of the lateral line, including a regeneration deficit associated with a change in behavior such as rheotaxis or escape reactions [72]. Exposure to 5 mg/L cadmium for 2 days severely damaged the nervous system of sea bass (*Dicentrarchus labrax*) [73].

4.2.2 Aminoglycosides

It was during the twenty-first century that the zebrafish lateral line was considered a model system for determining the toxicity of therapeutic drugs via hair cells. Williams and Holder [45] agreed that neomycin is toxic to lateral line hair cells. In addition, Harris et al. [17] showed that the response to neomycin is dose-dependent, identical in each neuromast, and that the sensitivity of hair cells depends on the age of the fish. Thus, in fish aged 4 days post fertilization (dpf), hair cells are less sensitive to neomycin than those of 5 dpf or more [74].

Van Trump et al. [75] also showed that hair cells in the lateral line are sensitive to damage caused by aminoglycosides. Research on the toxicity mechanisms of its molecules has shown that the swelling of mitochondria, loss of mitochondrial membrane potential, and the need for Bcl2 proteins associated with mitochondria suggest that aminoglycosides activate mitochondrial-dependent cell death pathways [76]. Also, the duration of hair cell death differs considerably for different aminoglycoside molecules, showing that toxicity pathways are initiated via the activation of different intracellular signaling cascades [46]. Actually, treatment with aminoglycosides lateral line of fish species has become a standard tool in behavioral studies designed to study lateral line function [77].

4.2.3 Cisplatin

The cisplatin toxicity study, which is a platinum-based chemotherapeutic agent used in the treatment of various tumors, has shown that there is a proportional relationship between the dose and the time of hair cell loss [66]. Thus, hair cells damage is related to intracellular accumulation of cisplatin as well as little work that reports cascading activation of cell death in lateral line hair cells treated with cisplatin. Molecules with antioxidant properties, such as N-acetyl L-cysteine and D-methionine, exhibit protective capacity for hair cells in zebrafish exposed to cisplatin, suggesting that cisplatin can induce oxidative stress pathways. These oxidative stress pathways have been implicated in the ototoxicity of cisplatin and aminoglycosides, suggesting some conservation of cell death mechanisms between different classes of ototoxic drugs [78]. Nuclear condensation and mitochondrial swelling are the consequences of apoptotic cell death [66].

4.2.4 Endocrine disruptors

Endocrine disruptors chemicals (EDCs) are natural or synthetic compounds found in the environment that disrupt the levels and distribution of endogenous hormones in exposed organisms [79], which can alter development and/or reproduction in humans and wildlife. Among the endocrine-disrupting compounds, we can cite natural hormones (17 beta-estradiol, E2) and synthetic [17α-ethinylestradiol (EE2)], pesticides [for example, dichlorodiphenyltrichloroethane (DDT)], polychlorinated biphenyls (PCB), bisphenol A (BPA), phthalates, flavonoids and polycyclic musks [80]. Zebrafish are used as a model organism to study endocrine disruptors and assess environmental risks [81]. The main mechanism of action triggered by these molecules is an agonist or antagonist interaction with ER. The concentration of vitellogenin (VTG), an egg yolk precursor protein, which is produced and secreted by the liver, absorbed by the ovary, and changed by developing eggs to form egg yolk, is the most common biomarker most widely used for EDC activity [81].

Bisphenol-A (BPA), polychlorinated biphenyls (PCBs), 17α-ethinylestradiol (EE2), and pesticides are widespread aquatic pollutants and can deeply affect the lateral line of zebrafish via disruption of endocrine system signaling. Bisphenol-A (BPA) is found in polycarbonate plastics and epoxy resins, as well as in the coatings of some cans. Researchers examined its impact on lateral line regeneration and found that it is toxic to hair cells in zebrafish larvae and that exposure delays regeneration [67]. BPA also attenuates hair cell regeneration after aminoglycoside damage, suggesting that BPA is toxic to supporting cells [71]. Hayashi et al. [67] found in another study that PCB-95 had no effect on lateral line development or hair cell survival. Rather, BPA had a significant effect on the survival and regeneration of hair cells. BPA-induced hair cell loss is both dose-dependent and temporal. Experimental laboratory studies suggest that BPA kills hair cells via activation of oxidative stress pathways, similar to previous reports of BPA toxicity in other tissues. In addition, Hayashi et al. [67] observed that hair cells killed with neomycin did not regenerate normally when BPA was present, suggesting that BPA in aquatic environments could interfere with innate regenerative responses in fish.

In other studies, Nasri et al. [68] examined the effect of the pesticide A6 derived from naturally-occurring α-terthienyl, and structurally related to the endocrine-disrupting pesticides anilinopyrimidines, on living zebrafish larvae. Results show that A6 decreases larval survival and affects central neurons at micromolar concentrations. In the lateral line system, researchers found that A6 alters the axonal and sensory cell regeneration at nanomolar concentrations. In addition, A6 has accumulated in lateral line neurons and hair cells. In addition, the examination of 17α-ethinylestradiol (EE2) effects at pico- to nano-molar concentrations on early nervous system development of zebrafish larvae showed that EE2 disrupted axonal nerve regeneration and hair cell regeneration. Upregulation of gene expression of estrogen and progesterone receptors has been recorded. In contrast, downregulation of the tyrosine hydroxylase, involved in the synthesis of neurotransmitters, occurred concomitant with diminution of proliferating cells. Collectively, EE2 modifies nervous system development, both centrally and peripherally, with negative effects on regeneration and swimming behavior [69].

5. Conclusions

The lateral line is a sensory system utilized at a variety of aquatic vertebrates especially in zebrafish, to detect changes in surrounding water flow. This sense, which utilizes mechanotransduction, mediates a wide variety of behaviors from predator detection to schooling. Its position on the body surface allows experimental laboratory studies. The regeneration of functional mechanosensory cells after damage or following pollutants exposure offers the potential to uncover processes involved in the maintenance, proliferation, and differentiation of sensory precursors. The latest research has approved that estrogen receptors are involved in the control of lateral line development and regeneration. These results support the use of this sensory system as a target for research on environmental estrogenic endocrine disruptors.

Conflict of interest

The authors declare no conflict of interest.

Author details

Ahmed Nasri[1,2]*, Patricia Aïssa[1], Ezzeddine Mahmoudi[1], Hamouda Beyrem[1] and Véronique Perrier[2]

1 Faculty of Sciences of Bizerta (FSB), Laboratory of Environment Biomonitoring, University of Carthage, Bizerta, Tunisia

2 U1198 MMDN (Molecular Mechanisms of Neurodegenerative Diseases), Inserm (National Institute for Health and Medical Research), MGX (Montpellier GenomiX), BioCampus, University of Montpellier, Montpellier, France

*Address all correspondence to: a7mednas@gmail.com; ahmed.nasri@fsb.u-carthage.tn

References

[1] McLachlan JA. Environmental signaling: What embryos and evolution teach us about endocrine disrupting chemicals? Endocrine Reviews. 2001;**22**(3):319-341

[2] Hertz R. Accidental ingestion of estrogens by children. Pediatrics. 1958;**21**(2):203-206

[3] Kavlock RJ et al. Research needs for the risk assessment of health and environmental effects of endocrine disruptors: A report of the US EPA-sponsored workshop. Environmental Health Perspectives. 1996;**104**: 715-740

[4] Birnbaum LS. State of the science of endocrine disruptors. 2013:a107-a107. DOI: 10.1289/ehp.1306695

[5] Sumpter JP. Xenoendocrine disrupters - Environmental impacts. Toxicology Letters. 1998;**103**:337-342

[6] Vos JG et al. Health effects of endocrine-disrupting chemicals on wildlife, with special reference to the European situation. Critical Reviews in Toxicology. 2000;**30**(1):71-133

[7] Jobling S et al. Widespread sexual disruption in wildfish. Environmental Science and Technology. 1998;**32**(17): 2498-2506

[8] Nash JP et al. Long-term exposure to environmental concentrations of the pharmaceutical ethinylestradiol causes reproductive failure in fish. Environmental Health Perspectives. 2004;**112**(17):1725-1733

[9] Mills LJ, Chichester C. Review of evidence: are endocrine-disrupting chemicals in the aquatic environment impacting fish populations ? Science of the Total Environment. 2005;**343**(1-3): 1-34

[10] Van Trump WJ, McHenry MJ. The morphology and mechanical sensitivity of lateral line receptors in zebrafish larvae (*Danio rerio*). Journal of Experimental Biology. 2008;**211**(13): 2105-2115

[11] Ghysen A, Dambly-Chaudiere C. Development of the zebrafish lateral line. Current Opinion in Neurobiology. 2004;**14**(1):67-73

[12] Nicolson T. The genetics of hearing and balance in zebrafish. Annual Review of Genetics. 2005;**39**:9-22

[13] Raible DW, Kruse GJ. Organization of the lateral line system in embryonic zebrafish. Journal of Comparative Neurology. 2000;**421**(2):189-198

[14] Gompel N et al. Pattern formation in the lateral line of zebrafish. Mechanisms of Development. 2001;**105**:69-77

[15] Chitnis AB, Nogare DD, Matsuda M. Building the posterior lateral line system in zebrafish. Developmental Neurobiology. 2012;**72**:234-255

[16] Wibowo I et al. Compartmentalized Notch signaling sustains epithelial mirror symmetry. Development. 2011;**138**:1143-1152

[17] Harris JA et al. Neomycin-induced hair cell death and rapid regeneration in the lateral line of zebrafish (*Danio rerio*). Journal of the Association for Research in Otolaryngology. 2003;**4**(2): 219-234

[18] Haas P, Gilmour D. Chemokine signaling mediates selforganizing tissue migration in the zebrafish lateral line. Developmental Cell. 2006;**10**(5):673-680

[19] Dambly-Chaudière C, Cubedo N, Ghysen A. Control of cell migration in the development of the posterior lateral

line: Antagonistic interactions between the chemokine receptors CXCR4 and CXCR7/RDC1. BMC Developmental Biology. 2007;7:23

[20] Venkiteswaran G et al. Generation and dynamics of an endogenous, self-generated signaling gradient across a migrating tissue. Cell. 2013;**155**(3): 674-687

[21] Xu H et al. Gb1 controls collective cell migration by regulating the protrusive activity of leader cells in the posterior lateral line primordium. Developmental Biology. 2014;**385**(2): 316-327

[22] Laguerre L, Ghysen A, Dambly-Chaudière C. Mitotic patterns in the migrating lateral line cells of zebrafish embryos. Developmental Dynamics. 2009;**238**(5):1042-1051

[23] Gamba L et al. Lef1 controls patterning and proliferation in the posterior lateral line system of zebrafish. Developmental Dynamics. 2010;**239**(12):3163-3171

[24] Matsuda M et al. Lef1 regulates Dusp6 to influence neuromast formation and spacing in the zebrafish posterior lateral line primordium. Development (Cambridge, England). 2013;**140**(11):2387-2397

[25] Aman A, Piotrowski T. Supplemental data Wnt/b-catenin and fgf signaling control collective cell migration by restricting chemokine receptor expression. Gene Expression. 2008;**15**(5):749-761

[26] Lecaudey V et al. Dynamic Fgf signaling couples morphogenesis and migration in the zebrafish lateral line primordium. Development (Cambridge, England). 2008;**135**:2695-2705

[27] Galanternik MV, Kramer KL, Piotrowski T. Heparan sulfate proteoglycans regulate Fgf signaling and cell polarity during collective cell migration. Cell Reports. 2015;**10**(3): 414-428

[28] Yanicostas C et al. Essential requirement for zebrafish anosmin-1a in the migration of the posterior lateral line primordium. Developmental Biology. 2008;**320**(2):469-479

[29] Hardelin JP, Dodé C, et al. The complex genetics of Kallmann syndrome: KAL1, FGFR1, FGF8, PROKR2, PROK2. Sexual Development. 2008;**2**(4-5):181-193

[30] Menuet A et al. Molecular characterization of three estrogen receptor forms in zebrafish: Binding characteristics, transactivation properties, and tissue distributions. Biology of Reproduction. 2002;**66**(6): 1881-1892

[31] Tingaud-Sequeira A et al. Expression patterns of three estrogen receptor genes during zebrafish (Danio rerio) development: Evidence for high expression in neuromasts. Gene Expression Patterns. 2004;**4**(5):561-568

[32] Froehlicher M et al. Estrogen receptor subtype β2 is involved in neuromast development in zebrafish (Danio rerio) larvae. Developmental Biology. 2009;**330**(1):32-43

[33] Celeghin A et al. The knockdown of the maternal estrogen receptor 2a (esr2a) mRNA affects embryo transcript contents and larval development in zebrafish. General and Comparative Endocrinology. 2011;**172**(1):120-129

[34] López-Muñoz A et al. Estrogen receptor 2b deficiency impairs the antiviral response of zebrafish. Developmental & Comparative Immunology. 2015;**53**(1):55-62

[35] O'Brien GS et al. Developmentally regulated impediments to skin reinnervation by injured peripheral

sensory axon terminals. Current Biology. 2009;**19**(24):2086-2090

[36] Villegas R et al. Dynamics of degeneration and regeneration in developing zebrafish peripheral axons reveals a requirement for extrinsic cell types. Neural Development. 2012;**7**(1): 1-14

[37] Gilley J, Coleman MP. Endogenous Nmnat2 is an essential survival factor for maintenance of healthy axons. PLOS Biology. 2010;**8**(1):e1000300

[38] Nguyen CT et al. The PAF1 complex component Leo1 is essential for cardiac and neural crest development in zebrafish. Developmental Biology. 2010;**341**(1):167-175

[39] Luo L, O'Leary DD. Axon retraction and degeneration in development and disease. Annual Review of Neuroscience. 2005;**28**:127-156

[40] Coleman MP, Freeman MR. Wallerian degeneration, wlds, and nmnat. Annual Review of Neuroscience. 2010;**33**:245-267

[41] White FV et al. Lipid metabolism during early stages of Wallerian degeneration in the rat sciatic nerve. Journal of Neurochemistry. 1989;**52**(4): 1085-1092

[42] Trapp BD, Hauer P, Lemke G. Axonal regulation of myelin protein mRNA levels in actively myelinating Schwann cells. Journal of Neuroscience. 1988;**8**(9):3515-3521

[43] Kaus S. The effect of aminoglycoside antibiotics on the lateral line organ of *Aplocheilus lineatus* (Cyprinodontidae). Acta Oto-Laryngologica. 1987;**103**(3-4): 291-298

[44] Song J, Yan HY, Popper AN. Damage and recovery of hair cells in fish canal (but not superficial) neuromasts after gentamicin exposure. Hearing Research. 1995;**91**(1):63-71

[45] Williams JA, Holder N. Cell turnover in neuromasts of zebrafish larvae. Hearing Research. 2000;**143**(1):171-181

[46] Coffin AB et al. Extracellular divalent cations modulate aminoglycoside-induced hair cell death in the zebrafish lateral line. Hearing Research. 2009;**253**(1):42-51

[47] Metcalfe WK, Kimmel CB, Schabtach E. Anatomy of the posterior lateral line system in young larvae of the zebrafish. Journal of Comparative Neurology. 1985;**233**(3):377-389

[48] Rubel EW, Furrer SA, Stone JS. A brief history of hair cell regeneration research and speculations on the future. Hearing Research. 2013;**297**:42-51

[49] Sun H, Lin CH, Smith ME. Growth hormone promotes hair cell regeneration in the zebrafish (Danio rerio) inner ear following acoustic trauma. PLoS One. 2011;**6**(11):e28372

[50] Faucher K et al. Damage and functional recovery of the Atlantic cod (*Gadus morhua*) inner ear hair cells following local injection of gentamicin. International Journal of Audiology. 2009;**48**(7):456-464

[51] Hernandez PP et al. Sub-lethal concentrations of waterborne copper are toxic to lateral line neuromasts in zebrafish (Danio rerio). Hearing Research. 2006;**213**(1):1-10

[52] Ma EY, Rubel EW, Raible DW. Notch signaling regulates the extent of hair cell regeneration in the zebrafish lateral line. Journal of Neuroscience. 2008;**28**(9):2261-2273

[53] Hernández PP et al. Regeneration in zebrafish lateral line neuromasts: Expression of the neural progenitor cell marker sox2 and proliferation-dependent and-independent mechanisms of hair cell renewal. Developmental Neurobiology. 2007;**67**(5):637-654

[54] Schuck JB et al. Transcriptomic analysis of the zebrafish inner ear points to growth hormone mediated regeneration following acoustic trauma. BMC Neuroscience. 2011;**12**(1):88

[55] Liang J et al. The stat3/socs3a pathway is a key regulator of hair cell regeneration in zebrafish stat3/socs3a pathway: Regulator of hair cell regeneration. Journal of Neuroscience. 2012;**32**(31):10662-10673

[56] Hertzano R et al. Transcription profiling of inner ears from Pou4f3 ddl/ddl identifies Gfi1 as a target of the Pou4f3 deafness gene. Human Molecular Genetics. 2004;**13**(18):2143-2153

[57] Namdaran P et al. Identification of modulators of hair cell regeneration in the zebrafish lateral line. Journal of Neuroscience. 2012;**32**(10):3516-3528

[58] Head JR et al. Activation of canonical Wnt/β-catenin signaling stimulates proliferation in neuromasts in the zebrafish posterior lateral line. Developmental Dynamics. 2013;**242**(7):832-846

[59] Jacques BE et al. The role of Wnt/β-catenin signaling in proliferation and regeneration of the developing basilar papilla and lateral line. Developmental Neurobiology. 2014;**74**(4):438-456

[60] Wada H et al. Wnt/Dkk negative feedback regulates sensory organ size in zebrafish. Current Biology. 2013;**23**(16):1559-1565

[61] Jiang L et al. Gene-expression analysis of hair cell regeneration in the zebrafish lateral line. Proceedings of the National Academy of Sciences. 2014;**111**(14):E1383-E1392

[62] Millimaki BB, Sweet EM, Riley BB. Sox2 is required for maintenance and regeneration, but not initial development, of hair cells in the zebrafish inner ear. Developmental Biology. 2010;**338**(2):262-269

[63] Lush ME, Piotrowski T. Sensory hair cell regeneration in the zebrafish lateral line. Developmental Dynamics. 2014;**243**(10):1187-1202

[64] Moon IS et al. Fucoidan promotes mechanosensory hair cell regeneration following amino glycoside-induced cell death. Hearing Research. 2011;**282**(1):236-242

[65] Walters BJ et al. Auditory hair cell-specific deletion of p27Kip1 in postnatal mice promotes cell-autonomous generation of new hair cells and normal hearing. Journal of Neuroscience. 2014;**34**(47):15751-15763

[66] Ou HC, Raible DW, Rubel EW. Cisplatin-induced hair cell loss in zebrafish (Danio rerio) lateral line. Hearing Research. 2007;**233**(1):46-53

[67] Hayashi L et al. The effect of the aquatic contaminants bisphenol-A and PCB-95 on the zebrafish lateral line. Neurotoxicology. 2015;**46**:125-136

[68] Nasri A et al. Neurotoxicity of a biopesticide analog on zebrafish larvae at nanomolar concentrations. International Journal of Molecular Sciences. 2016;**17**(12):2137

[69] Nasri A et al. Ethinylestradiol (EE2) residues from birth control pills impair nervous system development and swimming behavior of zebrafish larvae. Science of The Total Environment. 2021;**770**:145272

[70] Olivari FA, Hernández PP, Allende ML. Acute copper exposure induces oxidative stress and cell death in lateral line hair cells of zebrafish larvae. Brain Research. 2008;**1244**:1-12

[71] Mackenzie SM, Raible DW. Proliferative regeneration of zebrafish lateral line hair cells after different ototoxic insults. 2012;7(10):e47257

[72] Montgomery JC, Coombs S, Baker CF. The mechanosensory lateral

line system of the hypogean form of Astyanax fasciatus. The biology of hypogean fishes. Dordrecht: Springer; 2001. pp. 87-96

[73] Faucher K et al. Impact of acute cadmium exposure on the trunk lateral line neuromasts and consequences on the "C-start" response behaviour of the sea bass (Dicentrarchus labrax L.; Teleostei, Moronidae). Aquatic Toxicology. 2006;**76**(3-4):278-294

[74] Santos F et al. Lateral line hair cell maturation is a determinant of aminoglycoside susceptibility in zebrafish (Danio rerio). Hearing Research. 2006;**213**(1):25-33

[75] Van Trump WJ et al. Gentamicin is ototoxic to all hair cells in the fish lateral line system. Hearing Research. 2010;**261**(1):42-50

[76] Owens KN et al. Ultrastructural analysis of aminoglycoside-induced hair cell death in the zebrafish lateral line reveals an early mitochondrial response. Journal of Comparative Neurology. 2007;**502**(4):522-543

[77] Coombs S. Smart skins: Information processing by lateral line flow sensors. Autonomous Robots. 2001;**11**(3):255-261

[78] Rybak LP, Ramkumar V. Ototoxicity. Kidney International. 2007;**72**(8): 931-935

[79] Colborn T, Vom Saal FS, Soto AM. Developmental effects of endocrine-disrupting chemicals in wildlife and humans. Environmental Health Perspectives. 1993;**101**(5):378-384

[80] Gore AC et al. EDC-2: The Endocrine Society's second scientific statement on endocrine-disrupting chemicals. Endocrine Reviews. 2015;**36**(6):E1-E150

[81] Dang M et al. Long-term drug administration in the adult zebrafish using oral gavage for cancer preclinical studies. Disease Models & Mechanisms. 2016;**9**(7):811-820

Chapter 2

Role and Effect of Persistent Organic Pollutants to Our Environment and Wildlife

Nisha Gaur, Dhiraj Dutta, Aman Jaiswal, Rama Dubey and Dev Vrat Kamboj

Abstract

Persistent organic pollutants (POPs) are toxic substances composed of carbon-based chemical compounds and mixtures. In the recent times, persistent organic pollutants (POPs) came as a threat for the wildlife and environmental world. POPs are chemically stable, remain intact in the environment for long periods, are recalcitrant in nature, and are lipophilic in nature. Therefore, they accumulate in fatty tissue of living organisms and reside longer period of time finally affecting the human and wildlife. It is believed that approximately 90% of human beings are exposed to POPs from their diets that contain animal products. These foods lead to bioaccumulation in fat tissues that then cause health hazard. There are many studies such that its behavior in photocatalytic oxidation reactions are available; also many research studies are going to combat from these toxic substances. In this chapter, we will take you through how persistent organic pollutants are affecting our environment and wildlife and what are its roles.

Keywords: persistent organic pollutants, food chain, grasshopper, health, environment, Stockholm convention, global warming, marine

1. Introduction

As the name implies, persistent organic pollutants (POPs) are extremely persistent in the atmosphere, with a half-life of over a decade in the soil, sediments, air, and biota [1]. The importance of research into persistent organic pollutants is illustrated by the Stockholm Convention, adopted in 2001 by conference of Plenipotentiaries and came into force on May 17, 2004. POPs have now become the focus of different growing national and international concern as they show toxic effects on animal reproduction, development, and immunological function. Some national agencies are still not taking it seriously and call it differently as "Bio accumulative chemicals of concern" (BCCs) [2]. Only those compounds that get the extreme end of the distribution in degree of persistence, mobility, and toxicity will be ranked as POPs.

Chlorinated substances stay in the environment for a long time. With the introduction of electron capture detectors, several organochlorine pesticides such as dichlorodiphenyltrichloroethane (DDT), dichlorodiphenyldichloroethane (DDE), dieldrin, and toxaphene began to be detected [3]. These have been reported at such places where never been used before such as the earth's pole.

S. No.	Class	POPs
1	Subject to elimination of production and use (A)	Aldrin, hexachlorobenzene, Decabromodiphenyl ether, Endrin, Hexachlorobenzene, Beta hexachlorocyclohexane, Polychlorinated naphthalene's, Technical, endosulfan and its related isomers, Chlordane, Dicofol, Heptachlor, Hexabromocyclododecane, Hexachlorobutadiene, Lindane, Pentachlorophenol and its salts and esters, Perfluoroctanoic acid and its salts and PFOA related compounds, Tetrabromodiphenyl ether and pentabromodiphenyl ether, Chlordecone, Dieldrin, Hexabromodiphenyl ether, Alpha hexachlorocyclohexane, Mirex, Polychloriented biphenyls, Short-chain chlorinated paraffins, Toxaphene pentachlorobenzene, hexabromobiphenyl.
2	Restricted in production and use (B)	DDT, PFOS and PFOSF.
3	Unintentionally produced (C)	Pentachlorobenzene, hexachlorobenzene, Hexachlorobutadiene, PCDDs, polychlorinated Polychlorinated naphthalenes PCDFs, PCBs.
4	Chemicals under investigation	Dechlorane Plus, Methoxychlor, UV-328.

Table 1.
POPs recognized in Stockholm convention [4].

These bio-accumulate in the food chain (animals and humans), causing a slew of well-known health and environmental problems. These pollutants are causing great concern among scientists, governments, and nongovernmental organizations (NGOs). There are different types of POPs such as heptachlor, chlordane, aldrin, dieldrin, hexachlorobenzene, endrin, mirex, chlordecone, toxaphene, lindane, hexa- and penta-bromodiphenyl ethers (commercial octabromodiphenyl ether), tetra- and penta-bromodiphenyl ethers (commercial pentabromodiphenyl ether), hexabromobiphenyl, pentachlorobenzene, polychlorinated biphenyls (PCBs), α- and β-hexachlorocyclohexane, α- and β-endosulfans (technical endosulfan and its isomers), perfluorooctane sulfonic acid and its salts (PFOS), perfluorooctane sulfonyl fluoride (PFOSF), DDT, pentachlorobenzene, hexachlorobenzene, polychlorinated dibenzofurans (PCDFs), polychlorinated dibenzo-p-dioxins (PCDDs), polychlorinated biphenyls (PCBs), hexachlorobutadiene, chlorinated naphthalenes, pentachlorophenol hexabromocyclododecane (HBCD), short-chained chlorinated paraffins were recognized by United Nations Environment Program (UNEP), Montreal, International POPs Elimination Network (IPEN), and Stockholm Convention. These pollutants are classified into four categories based on their toxicity and are shown in **Table 1** [4].

2. POPs: a general aspect

It is important to discuss the aspects of POPs before coming to the effects of these to understand the notorious effects of POPs.

Persistent chemicals generally have higher concentrations because they usually present in the environment for longer time and get their steady state. If emissions decline to near-zero after a particular duration of use, the amount of chemical in the environment will decrease exponentially. The overall transformation rate constant in the environment determines the rate of elimination, and persistent substances are removed more slowly [1].

The impact of POPs can be well understood by its deposition process. This innovative method combines aerosol remote sensing data with POP aerosol phase observations. By this process, POPs impact remote oceanic areas, raising environmental concerns because of their toxicity and accumulation in aquatic food webs. Because POPs are semi-volatile chemicals that can exist in both the gas and aerosol phases, precipitation scavenging will deposit POPs in both dissolved and particulate forms [5]. Persistent organic pollutants separate into gas and aerosol phases once they are released and are subject to long-range atmospheric transport (LRAT). In the transport and fate of POPS at the regional and global scale, atmospheric depositional processes play a key role. Then, by adopting any of the three primary processes such as (1) dry deposition of particulate-bound pollutants, (2) diffusive gas exchange between the atmospheric boundary layer (ABL), and (3) the surface ocean and rain scavenging (either from gas or particulate phases), transport of semi-volatile organic compounds from the atmosphere to the ocean takes place [6]. Additionally, this can be concluded that the dry aerosol and gaseous deposition contribute to aquatic ecosystems pollutant burden and finally support POP accumulation in marine food web.

Climate change has already sparked a slew of environmental issues, and there is a direct link between pollutant emissions, dispersion, and toxicity and climate change. Both the IPCC report and the annual report of the United Nations Environment Program (UNEP) stressed the importance of paying attention to the problem of environmental pollution, particularly in light of global warming [7]. The series of POPs has already been discussed above, and the transport processes for persistent organic pollutants can be seen in **Figure 1**. POPs are transported globally in two ways: atmospheric circulation and ocean currents. POPs can exist in both gaseous and particle forms in the atmosphere. As a result, POPs in the atmosphere can be distributed globally via gaseous and particulate thanks to atmospheric circulation. POPs are more likely to be deposited on the land surface when the temperature drops while they will evaporate back into the atmosphere and migrate again, when the temperature rises.

POPs are transported globally in two ways: atmospheric circulation and ocean currents. POPs can exist in both gaseous and particle forms in the atmosphere. Hence, gaseous and particulate POPs in atmosphere can be globally spread with the help of atmospheric circulation.

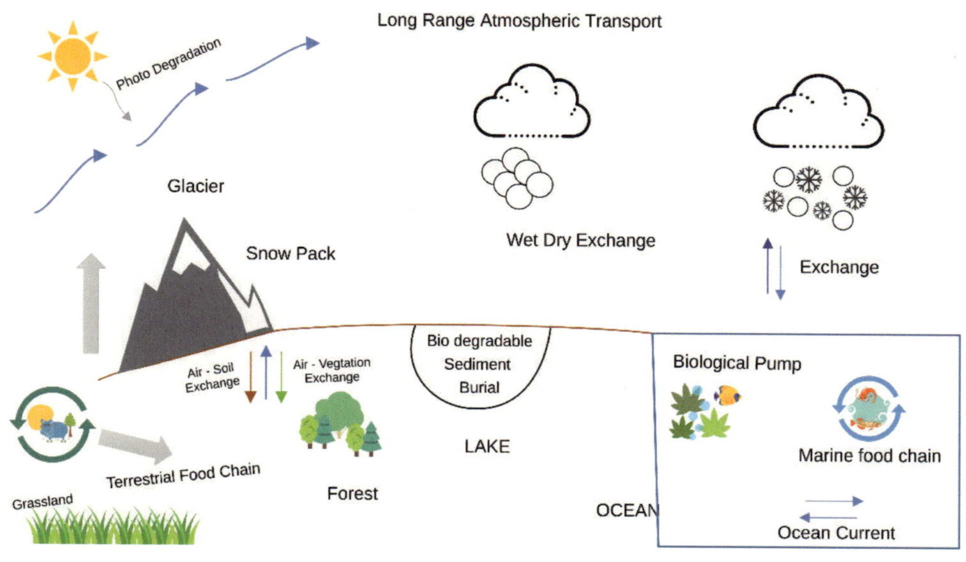

Figure 1.
Fate of persistent organic pollutants in the environment.

This cycle repeats itself, allowing POPs to be transported and deposited in far-flung locations. The grasshopper effect is what it is called. Furthermore, some POPs with a significantly higher solubility, such as hexachlorocyclohexanes (HCH) and perfluoro octane sulfonates (PFOS), can penetrate surface waters, feed into ocean currents, and travel around the world. Changes in climatic parameters such as temperature, wind speed, wind direction, and precipitation occur as a result of climate change. The intensity and pathways of POPs transported by air and ocean will undoubtedly change as these conditions change [8].

3. United Nations Stockholm convention

The text of the Stockholm Convention was adopted by the Conference of the Plenipotentiaries on May 22, 2001 and came into effect on May 17, 2004. In May 1995, the UNEP Governing Council recognized that persistent organic pollutants (POPs) pose serious and escalating dangers to human health and the environment and recommended that an international assessment process of an initial list of 12 POPs be performed in its resolution 18/32. The Intergovernmental Forum on Chemical Safety (IFCS) develops international action recommendations that must be considered by the UNEP Governing Council and the World Health Assembly by 1997. The Stockholm Convention's 12 initial POPs were divided into three categories.

- Pesticides: aldrin, dieldrin, endrin, heptachlor, chlordane, DDT, hexachlorobenzene, mirex, toxaphene;
- Industrial chemicals: hexachlorobenzene, polychlorinated biphenyls (PCBs); and
- By-products: hexachlorobenzene; polychlorinated dibenzo-p-dioxins and polychlorinated dibenzofurans (PCDD/PCDF), and PCBs.
- POPs are organic chemical compounds, that is, carbon-based, according to the Stockholm Convention. They have a unique set of physical and chemical qualities that, once released into the environment, allow them to:
- last an incredibly long time without deterioration
- As a result of natural processes including soil, water, and most significantly, air, they have become widely disseminated across the ecosystem;
- are lethal to both humans and wildlife and;
- accumulate in the fatty tissue of living organisms, including humans, and are present at higher levels in the food chain at larger quantities.

Another mechanism known as bioaccumulation concentrates POPs in living creatures [9]. The term "bioaccumulation" has two meanings: a) to define a dynamic process in which an element or compound is passively or actively taken up and concentrated within an organism; b) to indicate a currently high concentration as a result of prior accumulation activities [10].

4. Role and effect of persistent organic pollutants

Many developed countries have taken initiatives in recent years to restrict and limit the flow of PCBs into the atmosphere. The most important element in establishing these limits was a 1973 suggestion from the Organization for Economic Co-operation and Development (OECD) (WHO, 1976; IARC, 1978; OECD, 1982). Since then, the 24 OECD member countries have set manufacture, sales, importation, exportation, and use limitations, as well as a labeling system. POPs have severe effect on wildlife as well as on environment and hence to humans also. Reproductive impairment and malformations, increased risk of tumors, altered liver enzyme function are some of the many bad effects of POPs.

In **Figure 2**, it can be seen that POPs have an anthropogenic origin and are discharged into the air, water, and land, where they deposit in water and eventually enter the food chain through sediment. These are disseminated over the world by air and ocean currents and hence, travel a long distance. They subsequently penetrate atmospheric processes, air-water exchange, and cycles such as rain, snow, and dry particles, exposing even the most remote groups of humans and animals who rely on aquatic food.

4.1 Effect of POPs on environment

As of now, we know that POPs are a global hazard since POPs discharged into the environment can travel a considerable distance through the air and water via evaporation and redeposition from their initial source. The most important factor in the transportation of global POPs is the atmosphere. Because of the semi-volatile nature of the atmosphere, these chemicals are found in atmospheric gases. Once these POPs get encountered into the gases, they go under some other processes such that degradation, soil deposition, vegetation, bioaccumulation, sedimentation, and many more (**Figure 3**). Because these POPs are temperature-dependent, the Global Distillation Effect theory predicts that gas-phase contaminants will be transported from warm regions of the planet, such as tropical or temperate source areas, to colder, higher-latitude regions, affecting vapor pressure and Henry's constant,

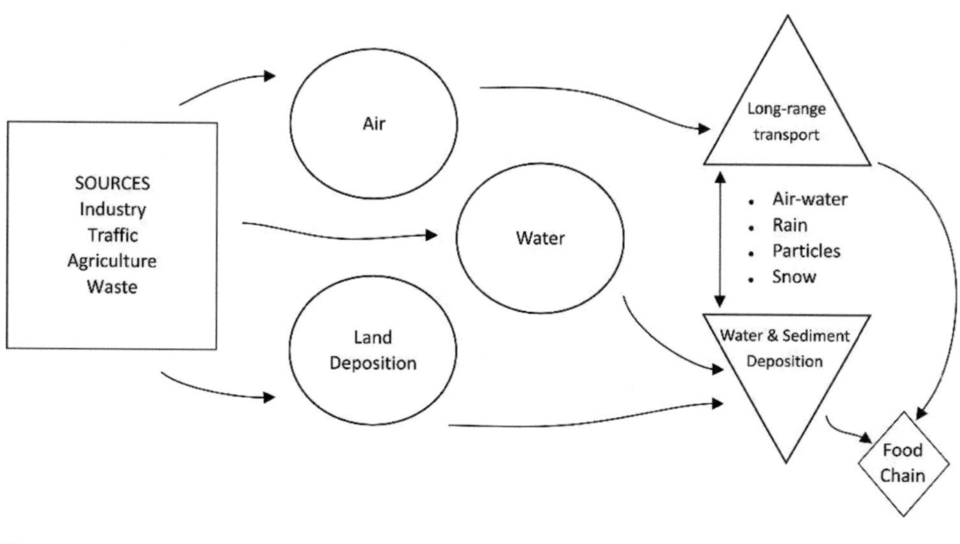

Figure 2.
POPs in the environment.

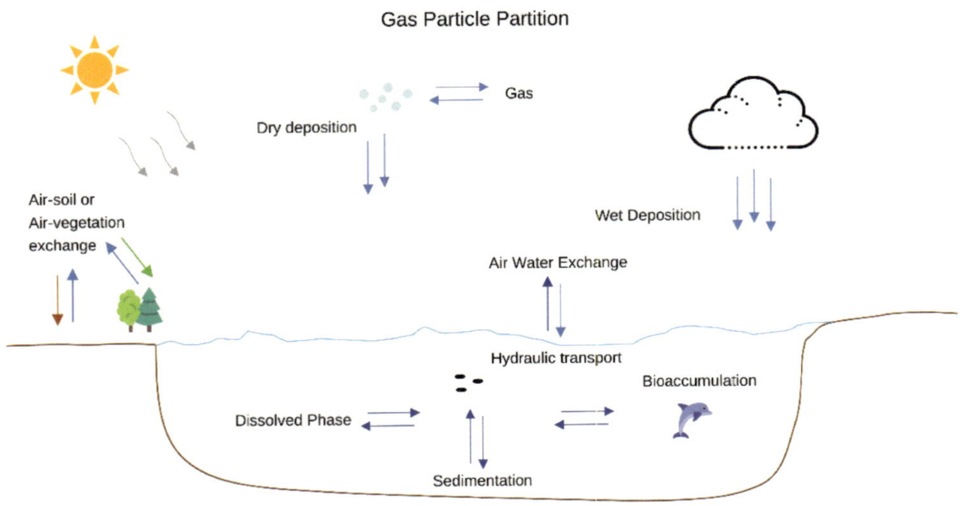

Figure 3.
Environmental progressions during long-range atmospheric transport of POPs.

resulting in condensation and accumulation of POPs in soil, vegetation, and other places, from where they can enter into the food chains [11].

Because reactivity, adsorption, and accumulation are all temperature-dependent activities, they can be influenced by climate change at any point along the transport and redistribution paths. Climate-change-related activities are predicted to modify POP exposure patterns for native and resident human populations in the long run. The majority of POPs have been produced and released in enormous quantities in the Northern Hemisphere, primarily by agriculture and industry. The emission pattern for most legacy POPs began around 1940–1950, according to published data on POPs. Following a significant increase in emissions, some countries banned or restricted the use of these compounds or found techniques to eliminate them as by-products, resulting in a period of reduction (about 1970). Variations in climate and ambient temperature have a direct impact on a number of deciding environmental elements as summarized in **Table 2** [12].

Several persistent organic pollutants (POPs), such as polychlorinated biphenyls (PCBs) and organochlorine pesticides (OCPs), have been widely utilized in Asian developing countries for industry, agriculture, and vector control throughout the last few decades. Pesticide usage in India is at 85,000 metric tons per year, with DDTs, HCHs, and malathion accounting for 70% of that [13]. Due to a lack of efficient facilities in Asian developing countries, substantial amounts of municipal trash from populous regions are directly thrown into open dumping sites with poor management. The public is concerned about the potential negative consequences for local communities and the environment as a result of this behavior. When recent rigorous research revealed increased human health risk from exposure to harmful substances such as dioxins and similar compounds, as well as heavy metals in these dumping sites, these worries became more serious [14].

Since these POPs are highly volatile and resistant to photolytic, biological, and chemical degradation, they were found at high concentrations around the globe, including open oceans, deserts, the Arctic, and the Antarctic. Many studies have looked at the spread of PCBs around the world. POP levels were found to be extremely high. Many studies have looked at the spread of PCBs around the world. Few studies have discovered high levels of some organochlorines in ocean, rainfall, and wild animals. The concentration of HCB in Antarctic fish was found to be comparable to that of North Sea fish in a study [15].

Parameters	Effects and indicators to be considered
Transport and distribution	Changing matrix composition Rapid/slow distribution Change of ocean currents, pH change, rising sea levels
Bioavailability	Changing bioavailability, new exposure routs invading species Changing regional food webs
Source and emissions	Increased emissions from secondary sources Emerging new primary sources
Environmental stability (persistence)	Changing thermal stability Changing radiation regime Changing adsorption properties Changing albedo
Toxicity/ecotoxicity	Changing toxicity potential Effects on immune response Exposure routs and profiles are changing Hormone and endocrine effects
Transformation and degradation	Changing degradation/transformation properties Changing microbiology

Table 2.
Influences of climate change on environmental processes related to transport and distribution of POPs [12].

4.2 Effect of POPs on wildlife

As we previously discussed, POPs are organic molecules with high lipid solubility, allowing them to stay in the environment for long periods of time, be transported substantial distances from their source, and bio-accumulate in food chains. They are deemed to have a danger of generating negative effects on human and wildlife health because of these traits.

Multiple exposure routes can expose wildlife species living in contaminated areas to complex combinations of pollutants. Wild species can be utilized as biomonitors of environmental pollution in a place since they have varied ecological, etiological, and physiological properties. Hazardous material exposure at various organizational levels also threatens the long-term viability of wildlife populations. As a result, pollution is currently considered one of the most serious threats to biological diversity. Exposure and effect biomarkers can be examined and integrated simultaneously to provide more information about probable toxicity pathways and ecosystem component health [16].

Marine wildlife numbers, particularly megafauna species, have dropped dramatically in recent decades, according to several research studies. For example, census studies in Eastern Australian seas show that loggerhead turtle numbers have plummeted by up to 86% during the 1970s. Similarly, in the last 90 years, the global population of dugongs appears to have fallen by at least 20%. (approximately three dugong generations). These and many more marine wildlife species have been added to the Red List of Threatened Species as a result of their critical population status, which spans from endangered to vulnerable to extinct, and are high conservation priorities [17].

Disease has been highlighted as a major contributor to the loss of marine wildlife populations, among the many concerns. Chronic sickness, which can lead to mortality, has been found to be on the rise in marine wildlife populations, reaching panzootic levels in some cases [18]. POPs are distributed in the marine environment as a result of subsequent transport mechanisms and source discharges. An increasing corpus of research examines the relationship between tissue loads of dominating POP groups and functional outcomes that have been degraded (**Table 3**).

Species	Target POPs	Functional parameters	Type of correlation	Reference
Gray seal	PBDEs	Circulating thyroid hormone	+	[19]
	PCBs, DDTs	TT3, FT3	—	[20]
California sea lion	PCBs	Serum vitamin A and T3	—	[21]
Beluga whale	PCBs	AhR and cyp1A1 expression	+	[22]
Bottlenose dolphin	PCBs	TT4, FT4, T3	—	[23]
	PCBs	Lymphocyte proliferation	—	[23]
	OH-PCBs	cyp1A1 expression	+	[24]
Common dolphin (male)	PCBs, DDTs	Blubber retinoids	+	[25]
Common dolphin (female)	PCBs, DDTs	Blubber retinoids	—	[25]
Polar bear	PCBs	Plasma cortisol	+	[26]
	OCPs	Plasma cortisol	—	[26]
Green sea turtle	p,p′-	Hematocrit	+	[27]
	DDE	Total blood proteins	—	[27]
	γ-HCH			

AhR: aryl hydrocarbon receptor; p,p′-DDE: 4,4′-dichlorodiphenyldichloroethylene; DDTs: dichlorodiphenyltrichloroethane; HCHs: hexachlorocyclohexanes- hexanes; OCPs: organochlorine pesticides; PCBs: polychlorinated biphenyls; PBDEs: polybrominated diphenyl ethers. TT3: total triiodothyrine; FT3: free triiodothyrine; TT4: total thyroxine; FT4: free thyroxine; T3: triiodothyrine; cyp1A1: cytochrome P450A1.

Table 3.
Summary of correlative studies associating functional health parameters with body burdens of persistent organic pollutants in marine wildlife.

Hydrophobic (water-hating) and lipophilic (fat-loving) compounds are the most common types of POPs. They bond firmly to solids, particularly organic matter, in both marine and terrestrial environments, avoiding the aqueous section. They also enter lipids more easily than the aqueous system inside cells and accumulate in fatty tissue. Chemicals are stored in fatty tissue, allowing them to remain in biota where metabolism is slow. As a result, POPs may move up the food chain. Under ambient temperatures, POPs tend to shift into the gas phase. As a result, they may volatilize from soils, plants, and aquatic systems into the air and migrate vast distances before being re-deposited due to their resistance to breakdown reactions in air [28].

POP loads can be passed from the mother to the child not only during placental development, but also through breast feeding in placental viviparous species, which are the only placental mammals. The importance of maternal transfer in terms of early exposure should not be ignored; various studies have shown that juvenile placental mammals acquire higher amounts of PCDD/Fs or PCBs via milk than they do from prenatal exposure during placental development. According to the studies, the offspring of pregnant rats administered six PCDD/Fs, including 2,3,7,8,-TCDD/F and the non-ortho PCBs 77, 126, and 169, got 7–28% of their doses lactationally and just 0.5–3% through the placenta. Fasting mothers can increase their children's dietary exposure to POPs from milk. Although the trend in female polar bear body burdens was not consistent—DDT and HCH decreased during fasting, while chlordane and PCBs increased—the ratio of plasma/adipose tissue and milk/adipose tissue OC concentrations did not change during the fast, indicating that POPs were probably at steady state among the various tagging systems [29].

All birds, most reptiles, most amphibians, and the rare monotreme mammals, for example, will deliver POPs to their eggs via maternal transfer. The direct transfer of the contaminant burden from the female to the eggs via the reallocation of the female's lipid storage is the principal source of POP exposure to the growing embryo. Both biological processes and chemical features of the pollutants induce the deposition of lipids and proteins (together with the POPs associated with the lipids and proteins) in the developing egg. The energy required for egg production can come from the female's older body reserves, her energy consumption during the egg formation period, or a combination of both [29].

4.3 Effect of POPs on human health

Among the many POPs that are abundant in our surroundings, a "black list" of POPs has been recognized under the diplomatic signature of the Stockholm Convention in 2001. Pesticides, such as aldrin, DDT, dieldrin, endrin, heptachlor, chlordane, mirex, and toxaphene; industrial compounds, such as hexachlorobenzene and polychlorinated biphenyls (PCBs); and other chemical by-products, such as polychlorinated dibenzodioxins (PCDDs) and polychlorinated dibenzofurans (PCDFs)—the general name "dioxins" is used for PCDDs and PCDFs. These POPs are known to be especially hazardous, having a high proclivity for biomagnification in the food chain, and have been linked to carcinogenic and endocrine disrupting effects in a variety of biota [9].

POP residues have been found in human adipose tissue from people living in a variety of countries, including Europe, Asia, Africa, and North America, for many years. Pesticide toxicity and persistence are beneficial for killing their target organisms, but they can cause difficulties for humans and the environment. Organochlorine pesticides (OCPs) have been found in the environment and, as a result, in the food chains of humans and wildlife since the early 1960s. Dichlorodiphenyltrichloroethane (DDT) and its derivatives have already been discovered in almost all environmental media and are the most common OCPs detected in human tissues, particularly adipose tissue. DDT is an organochlorine chemical initially synthesized in 1874 in Germany. The insecticidal effects of DDT were discovered in 1939, and commercial use began in 1945. DDT is dechlorinated in the human body to tetrachlorodiphenylethane (DDD), which is water-soluble and less hazardous to humans. Dichlorodiphenyldichloroethanes (DDEs) are another class of DDT derivatives that quickly accumulate in human adipose tissue and constitute a considerable health risk due to their long half-life. DDE can be acquired through DDT metabolism in the body or from intake of DDE-tainted foodstuffs [30].

While human data analysis has raised concerns, it has yet to produce conclusive evidence of causal relationships between low-level exposure to environmental chemicals, endocrine disrupting activities, and harmful human health impacts. All relevant data, including experimental animal data and wildlife observations, must be examined. The difficulties of correlating prenatal, postnatal, and childhood exposure to adult functioning are particularly concerning [31].

Reduced semen quality (i.e., reduced numbers, motility, and altered morphology of sperm), male reproductive tract abnormalities (e.g., hypospadias and cryptorchidism), altered sex ratio, endometriosis, precocious puberty, and early menarche) have been the focus of much of the human health concerns resulting from EDC exposure. A reduction in sperm counts has been recorded in a number of research studies in a lot of countries [32].

Increased rates of some hormone-related malignancies in many regions of the world are frequently cited as proof that EDC exposure has had negative health consequences. Increases in breast cancer and testicular cancer are particularly concerning. Several human epidemiological studies and experimental laboratory investigations have been undertaken to see if organochlorine pesticides are linked to an increased risk of breast cancer [31].

Organometals that bind to protein, particularly organomercurials; lipophilic contaminants such as dioxins, PCBs, polybrominated diphenyl ethers (PBDEs), and chlorinated pesticides; and persistent non-lipophilic compounds such as per fluorinated compounds used as repellents are all sources of concern for human health. Toxicity does not necessitate the persistence of a chemical. Many volatile organic chemicals, phthalates, and bisphenol A, which are present and leach from typical industrial products, are also found throughout the food supply and in the bodies of many of the world's populations. It has recently been discovered that exposure to certain chlorinated POPs increases the likelihood of developing type 2 diabetes, an insulin-related disease [33]. Surprisingly, this increased risk appears to occur at extremely low concentrations and does not appear to follow a linear dose–response curve [34]. DDE and hexachlorobenzene appear to have the strongest links, although PCBs also greatly enhance the risk [35]. Because type 2 diabetes is an insulin receptor disease, the processes underlying this link are unknown; however, it is most likely the result of gene induction. Obesity is frequently cited as a major risk factor for diabetes. Obese people who do not have high levels of POPs, on the other hand, do not have an increased risk of diabetes, according to certain research [33].

5. Conclusion

By now it is very much clear that how severe the POPs can be in future for the upcoming generations. It is not affecting only environment but also through environment to wildlife animals and directly or indirectly to human. Because of rising industrial use, persistent organic pollutants are becoming a major concern, causing their accumulation and persistence in living beings and the environment. Its exposure has numerous deadly implications for organs and tissues, including oxidative stress and cell death, due to several mechanisms. Several international plans have been developed to reduce the use of POPs and eliminate them completely; however, many developing countries around the world continue to ignore the Aarhus and Stockholm conventions. All of the class 1 and 2 POPs listed by these conventions must be banned. There are many agencies and NGOs working to accomplish their goals, but as humans, we must also recognize how damaging a single error can be. When scientific information is unknown, decision-making on the possible detrimental effects of chemical exposure is increasingly regulated by the precautionary principle, both internationally and nationally. The impact of endocrine disruptors on negative health consequences has yet to be proven conclusively. However, it is evident that the risk of endocrine disruptors is significant at specific times (preconception, pregnancy, and postpartum), and further research and development are needed to determine the health problems that should always be prioritized.

Conflict of interest

The authors declare no conflict of interest.

Author details

Nisha Gaur[1], Dhiraj Dutta[1], Aman Jaiswal[2], Rama Dubey[1*] and Dev Vrat Kamboj[1]

1 Defence Research Laboratory, DRDO, Tezpur, Assam, India

2 Indian Institute of Science Education and Research Mohali, Punjab, India

*Address all correspondence to: r_dubey172@rediffmail.com

References

[1] Jones KC, De Voogt P. Persistent organic pollutants (POPs): State of the science. Environmental Pollution. 1999;**100**(1):209-221

[2] Wania F, Mackay D. Tracking the Distribution of persistent organic pollutants. Environmental Science & Technology. 1996;**30**(9):390-396. DOI: 10.1016/s0926-3373(97)80026-4

[3] Muir DCG, Howard PH. Are there other persistent organic pollutants? A challenge for environmental chemists. Environmental Science & Technology. 2006;**40**(23):7157-7166. DOI: 10.1021/es061677a

[4] Alharbi OML, Basheer AA, Khattab RA, Ali I. Health and environmental effects of persistent organic pollutants. Journal of Molecular Liquids. 2018;**263**:442-453. DOI: 10.1016/j.molliq.2018.05.029

[5] Jurado E, Jaward F, Lohmann R, Jones KC, Simó R, Dachs J. Wet deposition of persistent organic pollutants to the global oceans. Environmental Science & Technology. 2005;**39**(8):2426-2435. DOI: 10.1021/es048599g

[6] Jurado E, Jaward FM, Lohmann R, Jones KC, Simó R, Dachs J. Atmospheric dry deposition of persistent organic pollutants to the Atlantic and inferences for the global oceans. Environmental Science & Technology. 2004;**38**(21):5505-5513. DOI: 10.1021/es049240v

[7] Climate Change 2013: The Physical Science Basis. Available from: https://www.ipcc.ch/report/ar5/wg1/ [Accessed July 22, 2021]

[8] Wang XP, Sun DC, Yao TD. Climate change and global cycling of persistent organic pollutants: A critical review. Science China Earth Sciences. 2016;**59**(10):1899-1911. DOI: 10.1007/s11430-016-5073-0

[9] Stockholm Convention. Available from: http://chm.pops.int/Implementation/ProgrammeofWork/tabid/6247/Default.aspx [Accessed July 23, 2021]

[10] Streit B. Fachbereich Biologie, University of Frankfurt, Siesmayerstrasse 70, D-6000 Frankfurt Am Main (Federal Republic of Germany). Experientia. 1992;**48**:955-970

[11] Fernández P, Grimalt JO. On the global distribution of persistent organic pollutants. Chimia (Aarau). 2003; **57**(9):514-521. DOI: 10.2533/000942903777679000

[12] Kallenborn R, Hung H, Brorström-Lundén E. Atmospheric Long-Range Transport of Persistent Organic Pollutants (POPs) into Polar Regions. Comprehensive Analytical Chemistry. Vol. 67. Elsevier; 2015. DOI: 10.1016/B978-0-444-63299-9.00013-2

[13] Minh NH, Minh TB, Kajiwara N, Kunisue T, Subramanian A, Iwata H, et al. Contamination by persistent organic pollutants in dumping sites of Asian developing countries: Implication of emerging pollution sources. Archives of Environmental Contamination and Toxicology. 2006;**50**(4):474-481. DOI: 10.1007/s00244-005-1087-3

[14] Agusa T, Kuntio T, Nakashima E, Minh TB, Tanabe S, Subramanian A, et al. Preliminary studies on trace element contamination in dumping sites of municipal wastes in India and Vietnam. Journal de Physique IV. 2003;**107**(1):21-24

[15] Jacob J. A review of the accumulation and distribution of persistent organic pollutants in the environment. International Journal of Bioscience, Biochemistry and Bioinformatics. 2013;**3**(6):657-661. DOI: 10.7763/ijbbb.2013.v3.297

[16] González-Mille DJ, Ilizaliturri-Hernández CA, Espinosa-Reyes G, Cruz-Santiago O, Cuevas-Díaz MDC, Martín Del Campo CC, et al. DNA damage in different wildlife species exposed to persistent organic pollutants (POPs) from the delta of the Coatzacoalcos River, Mexico. Ecotoxicology and Environmental Safety. 2018;**2019**(180):403-411. DOI: 10.1016/j.ecoenv.2019.05.030

[17] Jin L, Gaus C, Escher BI. Bioanalytical Approaches to Understanding Toxicological Implications of Mixtures of Persistent Organic Pollutants in Marine Wildlife. Comprehensive Analytical Chemistry. Vol. 67. Elsevier; 2015. DOI: 10.1016/B978-0-444-63299-9.00002-8

[18] Cunningham, andrew A. A walk on the wild side—Emerging wildlife diseases. BMJ. 2005;**331**(7527):1214-1215. DOI: 10.1136/bmj.331.7527.1214

[19] Bayen S, Gong Y, Chin HS, Lee HK, Leong YE, Obbard JP. Androgenic and Estrogenic response of green mussel extracts from Singapore's coastal environment using a human cell-based bioassay. Environmental Health Perspectives. 2004;**112**(15):1467-1471. DOI: 10.1289/ehp.6990

[20] Asawasinsopon R, Prapamontol T, Prakobvitayakit O, Vaneesorn Y, Mangklabruks A, Hock B. The association between organochlorine and thyroid hormone levels in cord serum: A study from Northern Thailand. Environment International. 2006;**32**(4):554-559. DOI: 10.1016/j.envint.2006.01.001

[21] Vanden Berghe M, Weijs L, Habran S, Das K, Bugli C, Pillet S, et al. Effects of polychlorobiphenyls, polybromodiphenylethers, organochlorine pesticides and their metabolites on Vitamin A status in lactating grey seals. Environmental Research. 2013;**120**:18-26. DOI: 10.1016/j.envres.2012.09.004

[22] Ostertag S. Bridging Knowledge Systems to Monitor Beluga Whale Health & Habitat Use in the Beaufort Sea. In: North. Contam. Progr. 21st Annu. Results Work. Canadian Science Publishing, Arctic Science; 2015. p. 2015

[23] Van Dolah FM, Neely MG, McGeorge LE, Balmer BC, Ylitalo GM, Zolman ES, et al. Seasonal variation in the skin transcriptome of common Bottlenose Dolphins (Tursiops Truncatus) from the Northern Gulf of Mexico. PLoS One. 2015;**10**(6):1-21. DOI: 10.1371/journal.pone.0130934

[24] Montie EW, Fair PA, Bossart GD, Mitchum GB, Houde M, Muir DCG, et al. Cytochrome P4501A1 expression, polychlorinated biphenyls and hydroxylated metabolites, and adipocyte size of Bottlenose Dolphins from the Southeast United States. Aquatic Toxicology. 2008;**86**(3):397-412. DOI: 10.1016/j.aquatox.2007.12.004

[25] Tornero V, Borrell A, Aguilar A, Forcada J, Lockyer C. Organochlorine contaminant and retinoid levels in Blubber of common Dolphins (Delphinus Delphis) off Northwestern Spain. Environmental Pollution. 2006;**140**(2):312-321. DOI: 10.1016/j.envpol.2005.07.006

[26] Sonne C, Dietz R, Born EW, Riget FF, Kirkegaard M, Hyldstrup L, et al. Is bone mineral composition disrupted by organochlorines in East Greenland Polar Bears (Ursus Maritimus)? Environmental Health Perspectives. 2004;**112**(17):1711-1716. DOI: 10.1289/ehp.7293

[27] Komoroske LM, Lewison RL, Seminoff JA, Deheyn DD, Dutton PH. Pollutants and the Health of Green Sea Turtles resident to an urbanized Estuary

in San Diego, CA. Chemosphere. 2011;**84**(5):544-552. DOI: 10.1016/j.chemosphere.2011.04.023

[28] Ashraf MA. Persistent organic pollutants (POPs): A global issue, a global challenge. Environmental Science and Pollution Research. 2017;**24**(5):4223-4227. DOI: 10.1007/s11356-015-5225-9

[29] de Solla SR. Exposure, bioaccumulation, metabolism and monitoring of persistent organic pollutants in terrestrial wildlife. The Handbook of Environmental Chemistry. 2016;**49**:203-252. DOI: 10.1007/698_2015_450

[30] Li QQ, Loganath A, Chong YS, Tan J, Obbard JP. Persistent organic pollutants and adverse health effects in humans. The Journal of Toxicology and Environmental Health, Part A, Current Issues. 2006;**69**(21):1987-2005. DOI: 10.1080/15287390600751447

[31] Damstra T. Potential effects of certain persistent organic pollutants and endocrine disrupting chemicals on the health of children. Journal of Toxicology: Clinical Toxicology. 2002;**40**(4):457-465. DOI: 10.1081/CLT-120006748

[32] Jørgensen N, Andersen AG, Eustache F, Irvine DS, Suominen J, Petersen JH, et al. Regional Differences in Semen Quality in Europe. Human Reproduction. 2001;**16**(5):1012-1019. DOI: 10.1093/humrep/16.5.1012

[33] Lee DH, Lee IK, Porta M, Steffes M, Jacobs DR. Relationship between serum concentrations of persistent organic pollutants and the prevalence of metabolic syndrome among non-diabetic adults: Results from the National Health and Nutrition Examination Survey 1999-2002. Diabetologia. 2007;**50**(9):1841-1851. DOI: 10.1007/s00125-007-0755-4

[34] Lee DH, Steffes MW, Sjödin A, Jones RS, Needham LL, Jacobs DR. Low dose of some persistent organic pollutants predicts type 2 diabetes: A nested case-control study. Environmental Health Perspectives. 2010;**118**(9):1235-1242. DOI: 10.1289/ehp.0901480

[35] Wang SL, Tsai PC, Yang CY, Guo YL. Increased risk of diabetes and polychlorinated biphenyls and dioxins A 24-year follow-up study of the Yucheng Cohort. Diabetes Care. 2008;**31**(8):1574-1579. DOI: 10.2337/dc07-2449

Chapter 3

Impact of PCBs, Furan and Dioxin on Hepatocarcinogenesis

Mohamed Helal, Sara Ghanem and Amany El-Sikaily

Abstract

Stockholm Convention defined polychlorinated biphenyls (PCBs) as a group of persistent organic pollutants (POPs) such as dioxin/furan, dichlorodiphenyltrichloroethane, polychlorinated biphenyls, polychlorinated dibenzo-p-dioxins, aldrin, polychlorinated dibenzofurans and organometallic compounds (such as organotin and organomercury) which share the same characteristics of being persistent, bioaccumulative and toxic and can travel long distance through various media. They have diverse health impacts with different underlined molecular mechanisms. Recently, PCBs were referred as potent carcinogens with persistent existence in the environment. As the liver is the organ of detoxification, it is the major target organ for toxic effects induced by environmental contaminants, including PCBs. PCBs, furan and dioxin exert their hepatocarcinogenic effect through different mechanisms such as induction of oxidative stress, an increase of reactive oxygen species (ROS), mutagenic induction to oncogenes and epigenetic alteration to hepatic cells. In this chapter, we will provide an updated overview about PCBs, furan and dioxins, their impact on liver cancer initiation and progression on various *in vivo* and *in vitro* systems and its underlined molecular mechanisms. Also, a special emphasis will be directed to highlight zebrafish as *in vivo* model system to analyse the hepatocarcinogenic effect of these pollutants.

Keywords: polychlorinated biphenyls, Dioxin, Furan, hepatocarcinogenesis, zebrafish

1. Introduction

Polychlorinated biphenyls (PCBs) are related to a larger group of environmental pollutants named persistent organic pollutants (POPs) with intense ecological burden and toxicological problems [1, 2]. PCBs are halogenated aromatic hydrocarbons with special chemical formulas encompassing 209 congeners [3]. Based on the number of chlorine atoms and their location at the biphenyl, PCBs can be divided into lower and higher chlorinated congeners. Lower and higher chlorinated PCBs have different bioaccumulation rates [4]. PCBs are made of two cis-carbon rings linked by a single carbon bond and biphenyl molecule. Each PCB molecule consists of 12 carbon atoms alongside chlorine atoms substituted for hydrogen ones at any of 10 possible positions. Hence, theoretically, 209 possible PCB congeners can be found [5]. Due to their chemical characteristics, they have been used extensively in the industry as electric insulators, plasticizers, heat exchange and hydraulic fluids. The main problem of PCBs lies in their persistent nature, high degree of lipophilicity, slow transformation rate and low environmental degradation, which made them associated with a broad spectrum of human diseases such as reproductive, immunological and carcinogenic [3, 6–8]. In 1987, International Agency for Research on Cancer and in 1996, the US environmental Protection Agency reported that PCBs are carcinogen in laboratory and wild animals and a possible carcinogen in humans [9–15].

Absorption of PCBs is feasible through ingestion, inhalation and dermal routes. The highest exposure pattern occurs through inhalation and skin absorption [16]. Once inside the body PCBs are transported to the liver. The main target of PCBs metabolism is the liver by the action of hepatic cytochrome p450 (CYP450). According to their chemical structure, PCBs can bind to different receptors [17] such as aryl hydrocarbon receptor (AhR), constitutive androstane receptor (CAR) and pregnane X receptor (PXR) [3]. Therefore, research was focused to study the impact of PCBs on liver function alteration and carcinogenicity [18]. Based on PCBs' chemical structure and receptor affinity, they can be categorized into dioxin-like and non-dioxin-like PCBs. Among different PCBs, Aroclors 1016, 1242, 1254 and 1260 are the most produced and used PCBs in the US during the 1958–1977 period [19].

One of the earliest incidences of human direct PCB ingestion was reported in Japan in 1968 and known as Yusho accident (oil syndrome in Japanese). This incident affected more than 1800 people who ingested rice oil contaminated with kanechlor-400 [20]. The average intake of PCBs in Yusho patients was 633 mg. The PCB concentration in adipose tissue was 46–76 ppm. After 34 years, the titre of non-ortho and mono-ortho PCBs in the blood of those patients reached 320 and 76 pg./g lipid, respectively [21]. Another famous PCB toxicity outbreak was reported a few years later after the Yusho incident in 1979 in Taiwan, known as the Yucheng incident (oil syndrome in Chinese). In this incident, around 2000 individuals consumed the same rice oil contaminated with Kanechlor-500 from one store [22]. Later, they developed skin problems and various health diseases. The concentration of PCBs in blood ranged from 3 to 1156 ppb [23]. A meta-analysis study of Yusho and Yucheng incidents showed that most patients exhibit a high degree of mortality due to lung, liver and skin cancers in both men and women [24, 25].

Before 1996, different assumptions have been made regarding PCBs' carcinogenicity. Sometimes all PCBs were considered carcinogens. Other assumptions indicated that mixtures with high chlorine content are only carcinogenic [26]. The carcinogenic potential of different PCBs is attributed to PCBs potency, which is affected by environmental processes (partitioning, chemical reaction, transformation and preferential bioaccumulation). Partitioning includes the fractionation of PCB mixture into different environmental compartments (air, water, sediment and soil). PCBs adsorption rate increases proportionally with their chlorine and organic content where low-chlorine content PCBs tend to be more volatile and hydrophilic and high-chlorine content is more persistent and lipophilic. Chemical transformation of PCBs in the environment occurs through the biodegradation process by the action of anaerobic bacteria in the sediments. These bacteria remove chlorines from *para* and *ortho* positions leading to a reduction of PCB toxicity [27]. PCBs bioaccumulation tends to concentrate in the adipose tissue with long-term stability and toxicity [28]. Due to the above-mentioned characteristics, EPA developed a new test approach to assess the cancer risk of different PCBs considering both environmental processes and PCBs toxicity [29]. Furthermore, Stecca *et al.* used different human cell lines (HuH6, HepG2 and DLD-1) to develop a suitable *in vitro* test battery to evaluate the cumulative effect of different PCBs mixtures. Their results showed that the best *in vitro* model to study toxicity is HepG2 and DLD-1 cell lines in terms of expression panel of several genes such as *AhR*, *AR*, *PXR*, *PPARγ*, *ERβ* and *THRα* (which showed the best representative expression upon PCBs mixture treatment) [30]. Most importantly, many of the genes identified by the Stecca group are also implicated in lipogenesis and lipid homeostasis in the liver indicating a significant role of PCBs on lipid metabolism in the liver [31].

Based on several reports from different agencies, PCBs were referred as possible human carcinogens [32]. This assumption is based on experimental data from rodents where PCBs treatment increased neoplasm formation in different rodent tissues.

Commercial PCB mixtures such as Aroclors 1016, 1242, 1254 and 1260 (dietary PCBs concentration ranged from 25 to 200 ppm over a period of 24-month treatment) were found to induce not only alteration in liver function tests (AST, ALT and GGT) but also liver tumours with bile duct carcinoma (cholangiocarcinoma) in rats (female tumour incidence higher than male rats) after long-term feeding regime [33]. In addition, rats exposed to commercial mixtures with 60% chlorine through a dietary lifetime regime developed benign liver tumours that eventually progressed to malignant ones [34]. Furthermore, a mixture of PCB126 and PCB153 caused a mild increase in neoplastic liver lesions in mice. This was accompanied by an up-regulation of Cyp1a1 and Cyp2b10 (RNA and protein level) [35]. The same promotion of liver carcinogenesis was observed in partially hepatectomized rats challenged with a single DEN dose and subsequent intraperitoneal injection of PCB105, 126 or 153 [36]. Moreover, rats receiving a single dose of DEN followed by intraperitoneal injection of PCB77 or PCB153 (150 μmol/kg) alongside selenium-enriched diets feeding developed hepatic neoplasm. Selenium administration enhances the carcinogenic induction of PCB77 more than that of PCB153 as the number of positive placental glutathione s0transferase (PGST+) hepatic regions was higher in the former than the latter, respectively [37].

Single PCB compounds have a preferential binding affinity to different receptors. For example, PCB126 binds to AhR while PCB153 binds to CAR. Rats treated with a single dose of DEN followed by PCB126, PCB153 or in combination developed hepatic neoplasms most profoundly in single PCBs treatment only (PCBs combination treatments showed antagonistic results on liver neoplasm formation) as indicated with positive GST-P liver areas [38]. Cultured mouse hepatocytes treated with PCB126 exhibited reduced hepatocyte glycogen content in a dose-dependent manner and suppressed forskolin-stimulated gluconeogenesis from lactate. Interestingly, glycogen treatment of cells restored PCB127 effects, indicating that PCB127 could affect the terminal players in the gluconeogenesis cycle. Finally, PCB126 could activate AhR and its downstream effector phosphoenolpyruvate carboxylase. This suggests a possible role of PCBs as energy metabolism disruptor agents [39]. Other studies showed that PCB153 could induce hepatocarcinoma through induction of NF-kB in mice (this was inhibited by deleting the p50 subunit of NF-kB) [40] or induce mutation in β-catenin (*Catnb*) [41] or *ras* [42] oncogenes as a promoter of tumorigenesis.

PCBs administration could interfere with metals accumulation in the liver and affect their transport and excretion through kidneys. Mice fed different concentrations of PCBs with Cadmium (Cd)-enriched diets showed a reduction of Cd concentration in the liver. Also, liver histology of those mice revealed a characterized centrilobular enlargement of hepatocytes, hepatic focal necrosis and clear cytological signs of malignancy than the control group [43]. On the other hand, female rats fed a diet enriched with high-dose Aroclor-1254 and Aroclor-1260 for 78 weeks developed initial iron accumulation in the liver by week 52, induced hepatocyte proliferation and eventually liver carcinoma by the 78th week, indicating that iron accumulation in the liver is an early sign of hepatic neoplasm transformation induced by PCBs [44].

Human exposure cohort studies were also conducted to monitor the pathological aspects of PCBs. Workers in capacitor factories exposed to Aroclors mixtures with 41–54% chlorine content had increased mortality rates from liver tumours (gall bladder and biliary tract) [45]. The same finding was reported in HCC Italian patients settled in areas highly polluted with PCBs [46, 47]. The burden of PCBs concentration in liver, lung and kidney tissues of Chinese cancer patients residing near e-waste disassembly sites was very high (257.9 to 455.1 ng g^{-1}), indicating a possible correlation between PCBs exposure and cancer incidence in those patients [48]. Another long-term cohort study was conducted in Germany in former PCB-exposed workers. The study linked the change in liver enzymes and morphology with PCB exposure level. There was a significant inverse connection between PCB concentration and ɣGT and a significant association between liver enlargement and PCBs level [49]. Another cohort study in the USA linked elevated levels of ortho-substituted PCBs and liver toxicant-associated steatohepatitis (TASH) in the former worker of PCB manufacturing complex. The authors reported that the increase in PCBs exposure was connected with an increase in liver disease burden, inflammation, steatohepatitis induction and hepatocyte apoptosis and fibrosis [50]. A large and extensive cohort study was conducted on 138,905 electricity workers exposed to insulating liquids of PCBs at five different electricity companies between the period of 1950 and 1986. Poisson regression was utilized to examine mortality of skin cancer (melanoma) and liver cancer in relation to PCBs exposure. Results showed that PCBs exposure was linked to melanoma development and in some workers hepatic cancers [51]. A controlled study was nested with two large prospective cohorts (one from Northern California Multiphasic Health Check-up (MHC) comprising 408 HCC cases and Norwegian Janus group comprising 84 HCC cases) from 1960 to 1980. Measuring 37 different congeners with GC-MS, the authors found that among measured congeners, PCBs (151, 170, 172,180,177 and 195) congeners were the highest with a concentration in HCC patients 1.3 to 1.4 ng/g lipid for the first group and 1.9 ng/g lipid for the second group, confirming a significant link between PCBs levels and HCC development [52].

PCBs can be indirectly accumulated in the human body through food chain by ingestion of aquatic animals contaminated with PCBs. For example, Delistraty study showed that PCBs titre in different aquatic animals in Columbia River, USA was significantly high. Sturgeon liver, whitefish fillet, carbs and smallmouth bass all showed significant high level of dioxin-like PCBs, non-dioxin like PCBs and total PCBs [53]. Another study showed that Bottlenose dolphin was stranded alive with high levels of different PCBs such as PCB 153,180, 187 and 138. Finally, large cell immunoblastic lymphoma was observed in the hepatic sinuses of these dolphins accompanied with liver enlargement. All previous studies indicate a direct correlation between carcinogenesis induction and levels of PCBs in those dolphins [9].

1.1 Possible hepatocarcinogenic mechanisms of PCBs

PCBs mode of action and the underlined molecular mechanisms of toxicity and carcinogenicity have not been deciphered so far [7]. Yet, studies on different animal models, *in vitro* cell lines and human cohort studies could give us a glimpse of the key molecular players responsible for different pathogenic outcomes.

2,3′4,4′,5-Pentachlorobiphenyl known as PCB118, one of the most persistent congener members, was found to promote hepatocellular carcinoma SMMC-7721 cell proliferation and glycolysis through AhR, which subsequently elevates the expression of pyruvate kinase M2 (PKM2) and stimulation of reactive oxygen species (ROS) production through nicotinamide adenine dinucleotide phosphate (NADPH). These effects were inhibited by treating cells with PKM2 shRNA and superoxide dismutase, respectively [6].

PCB126 (3,3′,4,4′,5-Pentachlorobiphenyl), a non-ortho-chlorinated congener, was found to increase the synthesis of ROS specifically. Treatment of HepG2 cells with this congener enhance their carcinogenicity by inducing an oxidative stress response that was underlined by activation of mitogen-activated protein kinases (extracellular signal-regulated kinase 1/2), p38, c-Jun phosphorylation, activating protein-1 (AP-1) and finally an expression of antioxidant-responsive element (ARE)-dependent genes [7]. In addition, Faust *et al*. demonstrated that rat progenitor liver cells (WB-F344) treated with PCB126 exhibit a differential transcriptional response over the treatment period. At 6-hour post-treatment (hpt), about 146 significant deregulated genes were identified under AhR direct targets. The number of deregulated genes was 658 and 968 after 24 and 72 hpt, respectively. The most identified genes through gene ontology analysis were affiliated to developmental, cell cycle, growth control and drug metabolism. The main targeted pathways were Wnt and TGF-β. Finally, they have also identified a novel target gene under the AhR signalling pathway such as *Fst*, *Btg2*, *Ctgf* and *Hbegf* [54]. AhR is an essential receptor controlling liver response to environmental toxicants. By using rat and human hepatocytes as *in vitro* cellular models to study PCBs toxicity, researchers found that rats fed Arochlor 1254 exhibit liver carcinoma through activation of AhR and downstream induction of *raf* effector in a MAPK-dependent pathway [55].

PCB47, 49, 52, 77 and 153 have a tumour promoting activity [32]. Also, some PCBs induce liver toxicity through induction of mixed function oxidases (phenobarbital, 3-methylcholanthrene) [56] and inhibition of anti-oxidant production such as PCB154, 155, 184 and 153 inhibit paraoxonase 1 (PON1) in treated rats [57].

Most PCBs mixtures with high chlorine content and their derived metabolites showed superior tumour-promoting characteristics. Yet, concern over low-chlorine content PCBs was raised after experiments showing that dihydroxy metabolites induce breast cancer by inducing oxidative DNA damage in breast cancer cells [58]. Another possible mechanism of PCBs carcinogenicity is their ability to suppress the immune system and cause endocrine disruption [59]. PCB104, 188 and their hydroxylated forms 4′-50, 4′-30, 4′-72, 4′-112 and 4′-121 disturb endocrine pathways in rainbow trout cultured hepatocytes and induced vitellogenin synthesis indicating altered liver physiology [60]. Human MCF-7 cells exposed to PCBs analogues showed a reduction in catechol-O-methyltransferase (COMT) activity on the transcriptional and translational level *via* the oestrogenic receptor. This could explain the PCB mode of liver tumour induction *via* modulation of oestrogen receptor response [61]. A comparative metabolomic study was conducted on rats fed a control diet and choline-deficient diet (as an inducer of liver non-alcoholic steatohepatitis) and subsequently exposed to PCB126. The addition of PCB127 promoted fatty liver development through dysregulation of glycerophospholipid metabolism, CoA biosynthesis pathway and glutathione metabolism. In addition to lipid metabolism disturbance, PCB127 down-regulated redox genes, and induced oxidative stress and mitochondrial dysfunction [62].

HepG2 cells co-treated with benzo-a-pyrene and different doses of Aroclor 1254 had a high degree of DNA damage (as indicated by DNA migration assay and formation of 8-hydroxy-2′-deoxyguanosine (8-OHdG)), oxidative stress and elevated CYP1A activity [63]. In another experiment, HepG2 cells exposed to various PCBs concentration (0.01-10μM) exhibit aggressive carcinogenic behavious underlined by pERK Tyr204 and pMdm2 Ser166 which attenuated P53 activity in those cells [64].

Dioxin-like PCBs such as PCB 77 and 81 were shown to have direct genotoxic effects on Chinese hamster V79-derived cell line by inducing micronuclei formation, and induced expression of CYP1A1, CYP2E1 and γ-H2AX protein (a marker of DNA double-strand breaks) [65, 66].

Another surprising finding of PCBs-induced hepatic carcinogenicity is their ability to inhibit intercellular community between liver cells. Mouse hepatoma cell line (Hepa1c1c7) treated with TCDD and different PCBs showed a rapid intercellular inhibition after 2 hrs. of treatment accompanied with AhR activation and induction of ethoxyresorufin O-deethylase (EROD) activity (an early marker of PCBs induced oxidative stress) [67]. Moreover, by using a quantitative polymerase chain reaction (qPCR) to quantify relative telomere length in lung and liver samples collected from rats treated with different PCBs (126, 153 and a mixture of them) showed larger relative telomere length, which is an early indication of euplastic or non-neoplastic pathogenic disease development [68].

2. Furan

Furan, a heterocyclic organic chemical, is considered as a human carcinogen and a liver toxicant in rodents [69]. It is found in a broad spectrum of common heat-treated and jarred foods in addition to tobacco smoke. It is also generated from numerous precursors such as amino acids, ascorbic acid and carbohydrates [70]. Infants received the highest furan exposure from ready-to-eat meals, while adults are exposed to furan by the dietary intake of coffee [69]. Furan is found mainly in the liver and is metabolized to the reactive metabolite, cis-but-2-ene-1,4-dialdehyde (BDA) through cytochrome P450 2E1 (CYP2E1). Reported studies have indicated that humans can convert furan to its reactive metabolite and cis-2-butene-1,4-dial (BDA), and consequently may be subjected to furan toxicity [71].

Being hepatotoxic, researchers [72] stated that furan is associated with cholangiofibrosis in rats and HCC & adenomas in mice. They also indicated that oxidative stress, alterations in gene expression, epigenetic modifications, inflammation and increased cell proliferation represent indirect mechanisms that are included in carcinogenesis. The carcinogenic effects of furan have been referred to as genotoxic and non-genotoxic modes of action. Epigenetic alterations are among the most important non-genotoxic alterations induced by furan since they are related to all other non-genotoxic events [69]. As a genotoxic furan could be linked to furan-into carcinogenicity, current human exposure levels to this hepatotoxicant may represent a risk to human health and required the necessity for its mitigation [73].

2.1 Possible hepatocarcinogenic mechanisms of furan

Metabolism of furan leads to the formation of protein adducts in the target organ. The first bioactivation step comprises the oxidation of cytochrome P450-catalysed of furan, which generates cis-2-butene-1,4-dial (BDA). BDA can react with lysine to form pyrrolin-2-one adducts [70]. This metabolite directly reacts with DNA nucleophiles and proteins [74]. It is also known as a bacterial mutagen in Ames assay strain TA104. According to metabolic studies, this reactive metabolite is formed *in vivo* [74]. BDA was found to react with glutathione (GSH) generating 2-(S-glutathionyl) butanedial (GSH-BDA), which reacts in turn with lysine forming GSH-BDA-lysine cross-links. Relative reactivity of these two intermediates was explored by the reaction of cytochrome *c* with BDA in the existence and absence of GSH [75]. Using MALDI-TOF mass spectrometry, BDA was found to react widely with cytochrome *c* forming adducts (which add 66 Da to the protein) according to pyrrolinone adducts formation. On the other hand, when GSH was added to the reaction, the overall extent of adduct formation was reduced. Briefly, the majority of adducts arose on lysine residues contributing to the carcinogenic hazard of furan [71]. By using liquid chromatography, tryptic peptides analysis clarified a cross-link between GSH-BDA and lysine 107 of histone H2B isolated from male F344 rats' liver exposed to carcinogenic doses of furan. This cross-link was detected before the identification of epigenetic changes and occurred at a lysine residue that is known as a target for epigenetic modifications and crucial for nucleosome stability [76].

Being a hepatocarcinogen in mice and rats, furan induced an enhancement for cytotoxic pathways represented by signalling of stress-activated protein kinase (SAPK) and death receptor (DR5 and TNF-alpha), and proliferation through extracellular signal-regulated kinases (ERKs) and TNF-α. In addition, NF-kappa B and c-Jun (genes essential for liver regeneration) were involved in response to furan [76]. Previous studies applying furan high doses revealed that it induced tumours at nearly 100% incidence at all doses [77]. Fraction of H-ras codon 61 CAA to AAA mutation was increased in liver tumours of furan-treated mice [78]. Besides, furan has a deleterious impact on the activity of crucial target enzymes included in ATP synthesis, glycolysis, redox regulation as well as b-oxidation in rat liver. After treatment with a high dose of furan, it was found that glyceraldehyde-3-phosphate dehydrogenase was significantly inhibited and observed some metabolic changes reliable to blockage of the glycolytic breakdown of glucose in the liver of the rat. Despite an increase in enoyl-CoA hydratase activity, an enhancement of ketone bodies production and a reduction in the activity of succinate dehydrogenase were recorded as a result of furan treatment. These enzymatic changes were linked to impairments occurring in cellular processes affecting the metabolic pathways and antioxidant defence and indicate mitochondrial dysfunction as a serious incident in furan toxicity [79]. Moreover, targets of putative protein of furan reactive metabolites induced functional damage of numerous individual proteins and interference with pathways, especially that of mitochondrial energy production, redox regulation, and protein folding. This damage represented critical targets of furan toxicity and can combine together to disturb cell homeostasis and cause the cell death of hepatocytes [80].

The liver is the main target organ affected by furan as indicated by serum biomarkers changes, change in liver weights and histological lesions after exposure to furan. Accordingly, a dose of 0.03 mg/kg bw of furan was proposed to be the non-detectable serious effect for hepatic toxicity [80]. In addition, Selmanoğlu *et al.* [81] revealed a significant increase of LDL levels, a significant decrease in ALT and ALP levels and insignificant changes in liver MDA levels, catalase activities and superoxide dismutase in the liver of rat groups treated with furan comparing with control groups. They also indicated a significant change in liver weights of furan-treated groups and observed hyperaemic blood vessels in their hepatic tissue under the light microscope. Histopathologically, multifocal hepatocellular necrosis intermingled with pigment-laden Kupffer cells and reactive leukocytes, oval-cell hyperplasia enhancement, hepatocyte mitoses increase and hepatocyte injury were also observed in livers from furan-treated mice as a result of furan induction [82]. Furthermore, furan-enhanced Ki-67 and PCNA expression in hepatic tissues increased the content of ROS in addition to indices of oxidative damage and decreased the TAC in the serum level of exposed rats. Finally, exposure to furan was found to be linked to changes in the mRNA expression pattern of intermediate filament proteins in hepatic tissues and promoted fibrosis and proliferation of hepatocytes in the liver [82].

In addition, analysis of liver rats treated with furan by Comet assays showed breaks in both strands of DNA, an increase in oxidized purines and pyrimidines at cancer bioassay dosage represented by a near-linear dose-responsive manner [83]. Consequently, these findings postulated that furan induces cancer mainly in rats' liver through a secondary genotoxic mechanism including oxidative stress, a down-regulation in the expression of apoptotic, cell-cycle checkpoints as well as DNA-repair genes accompanied by inflammation and cell proliferation dosage [83].

Furthermore, glutathione S-transferase placental form-positive (GST-P) foci are considered as preneoplastic lesions markers in the hepatocarcinogenic rats. Using reporter gene transgenic rats, it was found that furan rapidly induces GST-Pþ foci formation without reporter gene mutation after short exposure [84]. On the other hand, GST-P foci development is probably due to cell proliferation other than the genotoxic mode of action in furan-treated rats. Based on the close association between neoplastic hepatocytes and GST-P, Hibi *et al.* [85] postulated that cell proliferation following signal transduction other than the pathway of mitogen-activated protein kinase (MAPK)/ERK may contribute in the early stage of furan-induced hepatocarcinogenecity.

Cholangiofibrosis is defined as a physical anomaly that occurs before cholangiocarcinoma development in some rodents. Some reports explained that severely affected areas of the liver representing injury due to furan administration were extended into the portal and capsular parts, resulting in a rapid ductular cells proliferation that extended into the parenchyma accompanied by a subtype of liver fibroblasts. These ductules were differentiated into hepatocytes lacking fibroblasts or developed to form tortuous ductular structures replacing much of the parenchyma, leading to cholangiofibrosis [86]. Moreover, furan-induced cholangiocarcinomas were proposed to develop from cholangiofibrosis areas as a consequence of indirect and chronic damage to DNA through oxygen radicals joined with persistent proliferative signals, including loss of connexin 32, which acts to translate this DNA damage to fixed mutations [87].

2.1.1 Epigenetic alterations and the non-genotoxic mechanism of furan in liver

The carcinogenic effect of furan has been referred to as a genotoxic and non-genotoxic mechanism comprising epigenetic alterations in liver tissue [88]. Some reports postulated that furan carcinogenicity is caused by a non-genotoxic mechanism since it was not genotoxic in *in vivo* or *in vitro* micronucleus assay [89, 90]. Other studies indicated that BDA is not directly responsible for the effects of furan on mutational spectra *in vivo*. Therefore, an indirect mechanism of genotoxicity was hypothesized in which chronic toxicity was followed by inflammation and secondary cell proliferation that triggers the development of cancer in furan-exposed models [82].

In addition, epigenetic alterations involving DNA and microRNA (miRNA) methylation play a fundamental role in inducing furan carcinogenicity. It was indicated that DNA methylation changes and miRNA modulation followed by a DNA-damage response are the most pronounced alterations resulted from the use of 3-month furan treatment at a carcinogenic dose suggesting that non-genotoxic mechanisms are crucial for furan carcinogenicity [91]. It was found that gene-specific DNA methylation alterations have an essential role in the contribution of furan hepatotoxicity and hepatocarcinogenicity [88]. Other studies indicated that aberrations in microRNAs (miRNAs) expression are one of the non-genotoxic alterations induced by furan exposure, which highlighted the role of epigenetic impairments in the furan hepatotoxicity mechanism [69].

Moreover, Conti *et al.* [92] mentioned that epigenetic modifications which occurred in hepatotoxicity and carcinogenicity of furan are dose and time-dependent. They noted some epigenetic aberrations represented by DNA methylation, histone lysine acetylation and methylation, gene-specific methylation and alteration of chromatin-modifying genes expression in male Fisher rats treated with furan. Their findings indicated that sustained alterations in histone lysine acetylation (which is responsible for the ability of cells to maintain and control correctly the expression of genetic information) represent the adverse effects of furan induction. Some reports indicated that gene expression alterations resulted from furan exposure were irreversible [91]. Using whole-genome transcriptomic analysis, Tryndyak *et al.* [88] demonstrated differential gene expression alterations in liver lesions induced in male rats treated with furan. These alterations are essential in key pathways linked with the diverse aspects of liver pathology. Furthermore, it was noted that the continuous exposure to furan induced noticeable changes in the expression of miRNA represented by over-expression of hepatic miRNAs (miR-34a, miR-93, miR-200a, miR-200b and miR-224), and down-regulation of miR-375. In addition, hypermethylation of cytosine DNA and the lysine methylation of histone H3K9 and H3K27 at the MiR-375 genes were increased due to the reduction in miR-375 expression. It was revealed that the significant miR-375 inhibition was accompanied by toxicity and carcinogenicity of furan-induced liver leading to an up-regulation in Yes-associated protein 1 (YAP1), which is one of the principal events in liver carcinogenesis [69].

Since the mammalian genome is transcribed into mRNAs that code for protein and other non-coding RNA products [93]. Long non-coding-RNAs (lncRNA) are known as ncRNA species >200 nucleotides long, which represent significant epigenetic regulators of gene expression and are included in a wide spectrum of biological processes related to toxicology. Recio *et al.* [93] indicated that lncRNAs are transcriptional targets in the cytotoxic levels of furan exposure inducing cell proliferation. They also hypothesized that lncRNAs are considered as epigenetic biomarkers of carcinogenic exposure.

3. Dioxins

Dioxins are considered as representative toxic agents among persistent organic pollutants and a large family of halogenated aromatic hydrocarbons, which composed of tricyclic aromatic compounds [94]. These compounds are produced by industrial wastes and can accumulate in soil, sediments as well as food chains with long half-life of numerous years, affecting human health [95]. 2,3,7,8-Tetrachlorodibenzo -p-dioxin (TCDD) is a typical representative and the most toxic substance of dioxins, which exhibits systemic hepatotoxicity, carcinogenicity, immunotoxicity, teratogenicity, endocrine disruption and also affects pathology and physiology of human skin [96]. Being with four chlorine atoms in lateral positions, 2,3,7,8 Tetrachlorodibenzop-dioxin (TCDD) is the most biologically active isomer of dioxins [97]. It is a widespread and persistent pollutant in the environment originated from waste incineration or metal industries, plastics manufacturing and paper processing [98]. Moreover, it plays a significant role by binding to AhR for endocrine changes in experimental animals [99]. Besides, Türkez, Türkez *et al.* [100] suggested that oxidative stress has a crucial role for toxic effects of TCDD with AhR.

3.1 TCDD exposure and dioxin receptor

The aryl hydrocarbon receptor (AhR) is considered as a ligand-activated receptor which enables environmental pollutant toxicity like 2,3,7,8-tetrachlorodibenzo-p-dioxin (TCDD) [101]. It is also known as xenobiotic receptor or dioxin receptor and is a member of the basic helix-loop-helix/period AhR nuclear translocator single-minded family [102]. AhR translocates to the nucleus after binding to TCDD, dimerizes with AhR nuclear-translocator protein, binds to dioxin-responsive elements and up-regulates a series of genes expression that encode xenobiotic-metabolizing enzymes, such as cytochrome P450s (e.g. CYP1A1, CYP1A2), NAD(P)H quinone oxidoreductase as well as a form of UDPglucoronosyl-transferase-6 [103]. Even though AhR may serve as part of an adaptive chemical response, numerous studies reported that this dioxin receptor has important functions in liver, cell proliferation, cardiac development [104, 105] and the ubiquitin-proteasome system [106]. AhR plays a fundamental role in three biological aspects including xenobiotics metabolism, the toxic responses related to TCDD (dioxin) exposure and the vascular remodelling of the developing embryo. Using *Cre*-lox technology, Walisser *et al.* [107] examined the role of AhR signalling in hepatocytes and endothelial cells. They revealed that AhR signalling in hepatocytes is crucial to produce adaptive and toxic hepatic responses due to TCDD exposure.

Being a generally expressed ligand-dependent transcription factor, AhR mediates cellular responses to dioxins. Boutros *et al.* [108] demonstrated that AhR mediated all effects of dioxin on hepatic mRNA levels and revealed the alteration of 297 genes including many well-established AhR target genes due to dioxin exposure in mice liver. They also indicated that AhR genotype remodelled hepatic transcriptomes suggesting the existence of a basal AhR gene battery. Results of Boutros *et al.* [108] highlighted the fundamental role of this dioxin receptor in the liver tissue.

In response to dioxin, Kennedy *et al.* [109] also explained that the signal transduction pathways that mediate tumour promotion of liver by 2,3,7,8-tetrachlorodibenzo-*p*-dioxin are accomplished by the linked action of two receptor systems, the AhR and the receptors for the "IL-1-like" cytokines. However, Yamaguchi and Hankinson [110] indicated that TCDD might suppress the cell growth of liver cancer through numerous signalling pathways, mediated by AhR and its related co-factors. In addition, they found that the impact of TCDD was accomplished by gemcitabine (responsible for nuclear DNA damage in cancer cells), suggesting that their use as a combination may be considered as a suppressor of tumour cell growth *in vitro*.

TCDD can induce hepatic fibrosis through a sequential events of steatosis followed by steatohepatitis. Lee, Wada [102] investigated the role of AhR in liver steatosis in wild type and transgenic mice. They concluded that AhR activated in liver cells induced CD36 expression, enhanced the uptake of fatty acids and steatosis induciton [102].

Cytochrome P4502E1 (CYP2E1) mainly expressed in liver, is involved in the metabolic activation of carcinogens and hepatotoxins such as TCCD and CCl4. At post-transcriptional levels, CYP2E1 is induced and exerted mostly through mRNA and protein stabilization, while xenobiotic induction is found to be very limited at the transcriptional level [101]. Since the effect of xenobiotics on CYP2E1 liver, expression is of significant attention. Therefore, Mejia-Garcia *et al.* [101] studied the effect of TCDD on CYP2E1 liver of mouse and the impact on CCl4 that induced hepatotoxicity. They found that TCDD augmented levels of mRNA and protein in hepatic tissue of mouse CYP2E1 in an AhR-dependent manner and CYP2E1 was induced causing CCl4-induced hepatotoxicity.

Recent studies revealed that TCDD exposure had caused increased productions of lipid peroxidation, reactive oxygen species and histopathological injury in the liver of both rats and mice [111]. This exposure also enhances oxidative stress and diminishes the fluidity of hepatic membrane and glutathione (GSH) content, as well as imbalances the antioxidant enzymes in the liver [112, 113]. Moreover, an increase in the relative weight of the liver, a significant increase in all of the hepatic biomarker levels (glucose, cholesterol, triglycerides, AST, ALT and LDH) in the serum and a decrease of the antioxidant enzyme activities (catalase, glutathione peroxidase and superoxide dismutase) were observed under dioxin effect in hepatic tissue of rat [114]. Additionally, Bentli *et al.* [99] revealed that immunotoxicity associated with altered cytokine levels is among the other TCDD-induced toxicity prominent symptoms. Finally, Ciftci and Ozdemir [115] indicated that one of the main regulated pathways of TCDD toxicity is the elevated levels of the inflammatory cytokine. Using the real-time polymerase chain reaction (PCR), several studies indicated that heat shock proteins (mortalin, α-B-crystallin, Hsp105, Hsp27 and Hsp90s) and antioxidant enzymes (GST, SOD-3 and catalase) in livers of rats were induced suggesting protective mechanisms against 2,3,7,8-TCDD which induced hepatotoxicity [116]. Moreover, Czepiel *et al.* [117] mentioned that TCDD impaired the liver of rats and the activity of CYP1A1 in a dose-dependent manner. Parenchymal degeneration of hepatic lobules, hepatocytes vacuolation in prominent and peripheralized nuclei, hepatocellular hypertrophy and turgor of the vein in the centriacinar regions were also observed in rats' liver that received a high dose of dioxin [114]. TCDD also induces CYP1A1 activity by elevating the immunohistochemical reactivity of central areas of hepatic lobules located around the central vein in the rat liver [117].

4. Zebrafish models of PCBs-, furan- and dioxin-induced hepatocarcinogenesis

Viluksela and Pohjanvirta [118] reported that paternal exposure to TCDD was considered as the most effective congenator of dioxins in laboratory rodents and zebrafish as it can lower the reproductive performance and reduce the male/female ratio of offspring. Therefore, it will affect subsequent generations *via* both paternal and maternal germlines. These adverse effects have been accompanied by epigenetic alterations in sperm cells and/or placenta, including variations in methylation patterns of imprinted genes.

Besides, previous studies have demonstrated that dioxins broadly alter hepatic mRNA levels [119]. Unexpectedly, Boverhof, Burgoon [120] found that responses of mouse and rats to TCDD exposure revealed that rat and mouse responses diverge significantly through analysis of a limited portion of their transcriptome. Accordingly, it was suggested that both mice and rat models should be applied to detect the acute hepatotoxicity of xenobiotics [120].

Under fabp10a promoter, Zhang [121] also established a line of transgenic zebrafish line (LiPan) characterized with the expression of liver-specific red fluorescent protein (DsRed), which enables the observation of liver in live LiPan fry. They revealed that TCDD could significantly increase both liver red fluorescence and size in LiPan fry. Thus, LiPan transgenic fry offers a suitable and rapid hepatotoxicity assay *in vivo* that should be used to monitor the effect of environmental contaminants from the chemical mixture [121].

Furthermore, by using the inducible *kras* transgenic zebrafish model of hepatocarcinogenesis, Qiqi *et al.* have shown that PCB12 and TCDD alongside other environmental pollutants could accelerate HCC induction and inflammation of the liver [122]. Zebrafish is a very promising model of environmental and molecular toxicology. Further studies and more deep studies should be carried out using this model to provide more insightful information about carcinogenic potential and mechanisms of PCBs, furan and dioxins.

5. Conclusion

Environmental pollutants are a severe persistent burden, which cause a broad spectrum of health problems not only to aquatic animals but also to humans. Among them, PCBs, furan and dioxin are organic pollutants that were widely used in different applications before they were banned due to their carcinogenic potential. Different studies using different model animals and screening systems (*in vivo* and *in vitro*) indicated their correlation with liver tumour induction and promotion. In this chapter, we highlighted the collective and recent updates linking these groups of pollutants with the pathology of hepatocellular carcinoma. Although extensive research has been done, yet the exact potential and molecular mechanisms of these pollutants are to be discovered and deciphered.

Author details

Mohamed Helal*, Sara Ghanem and Amany El-Sikaily
National Institute of Oceanography and Fisheries (NIOF), Cairo, Egypt

*Address all correspondence to: m.helalf@gmail.com

References

[1] Dirinck E, Jorens PG, Covaci A, Geens T, Roosens L, Neels H, et al. Obesity and persistent organic pollutants: Possible obesogenic effect of organochlorine pesticides and polychlorinated biphenyls. Obesity (Silver Spring, Md). 2011;**19**(4):709-714

[2] Markowitz G, Rosner D. Monsanto, PCBs, and the creation of a "world-wide ecological problem". Journal of Public Health Policy. 2018;**39**(4):463-540

[3] Faroon O, Ruiz P. Polychlorinated biphenyls: New evidence from the last decade. Toxicology and Industrial Health. 2016;**32**(11):1825-1847

[4] Safe S. Toxicology, structure-function relationship, and human and environmental health impacts of polychlorinated biphenyls: Progress and problems. Environmental health perspectives. 1993;**100**:259-268

[5] Carpenter DO. Polychlorinated biphenyls (PCBs): Routes of exposure and effects on human health. Reviews on Environmental Health. 2006;**21**(1):1-23

[6] Liang W, Zhang Y, Song L, Li Z. 2,3′4,4′,5-Pentachlorobiphenyl induces hepatocellular carcinoma cell proliferation through pyruvate kinase M2-dependent glycolysis. Toxicology Letters. 2019;**313**:108-119

[7] Song MO, Freedman JH. Activation of mitogen activated protein kinases by PCB126 (3,3′,4,4′,5-pentachlorobiphenyl) in HepG2 cells. Toxicological Sciences. 2005;**84**(2):308-318

[8] Knerr S, Schrenk D. Carcinogenicity of "non-dioxinlike" polychlorinated biphenyls. Critical Reviews in Toxicology. 2006;**36**(9):663-694

[9] Jaber JR, Pérez J, Carballo M, Arbelo M, Espinosa de los Monteros A, Herráez P, et al. Hepatosplenic large cell immunoblastic lymphoma in a bottlenose dolphin (Tursiops truncatus) with high levels of polychlorinated biphenyl congeners. Journal of Comparative Pathology. 2005;**132**(2-3):242-247

[10] Program NT. Toxicology and carcinogenesis studies of 2,3,4,7,8-pentachlorodibenzofuran (PeCDF) (Cas No. 57117-31-4) in female Harlan Sprague-Dawley rats (gavage studies). National Toxicology Program Technical Report Series. 2006;**525**:1-198

[11] Program NT. Toxicology and carcinogenesis studies of a binary mixture of 3,3′,4,4′,5-pentachlorobiphenyl (PCB 126) (Cas No. 57465-28-8) and 2,2′,4,4′,5,5′-hexachlorobiphenyl (PCB 153) (CAS No. 35065-27-1) in female Harlan Sprague-Dawley rats (gavage studies). National Toxicology Program Technical Report Series. 2006;(530):1-258

[12] Program NT. Toxicology and carcinogenesis studies of a mixture of 2,3,7,8-tetrachlorodibenzo-p-dioxin (TCDD) (Cas No. 1746-01-6), 2,3,4,7,8-pentachlorodibenzofuran (PeCDF) (Cas No. 57117-31-4), and 3,3′,4,4′,5-pentachlorobiphenyl (PCB 126) (Cas No. 57465-28-8) in female Harlan Sprague-Dawley rats (gavage studies). National Toxicology Program Technical Report Series. 2006;(526):1-180

[13] Program NT. Toxicology and carcinogenesis studies of a binary mixture of 3,3′,4,4′,5-pentachlorobiphenyl (PCB 126) (Cas No. 57465-28-8) and 2,3′,4,4′,5-pentachlorobiphenyl (PCB 118) (Cas No. 31508-00-6) in female Harlan Sprague-Dawley rats (gavage studies). National Toxicology Program Technical Report Series. 2006;**531**:1-218

[14] Program NT. Toxicology and carcinogenesis studies of

2,3′,4,4′,5-pentachlorobiphenyl (PCB 118) (CAS No. 31508-00-6) in female harlan Sprague-Dawley rats (gavage studies). National Toxicology Program Technical Report Series. 2010;**559**:1-174

[15] N Program NT. TP toxicology and carcinogenesis studies of 3,3′,4,4′,5-pentachlorobiphenyl (PCB 126) (CAS No. 57465-28-8) in female Harlan Sprague-Dawley rats (Gavage Studies). National Toxicology Program Technical Report Series. 2006;**520**:4-246

[16] Bosetti C, Negri E, Fattore E, La Vecchia C. Occupational exposure to polychlorinated biphenyls and cancer risk. European Journal of Cancer Prevention: The Official Journal of the European Cancer Prevention Organisation (ECP). 2003;**12**(4):251-255

[17] Oliver JD, Roberts RA. Receptor-mediated hepatocarcinogenesis: Role of hepatocyte proliferation and apoptosis. Pharmacology & Toxicology. 2002;**91**(1):1-7

[18] Ludewig G, Robertson LW. Polychlorinated biphenyls (PCBs) as initiating agents in hepatocellular carcinoma. Cancer Letters. 2013;**334**(1):46-55

[19] Mayes BA, McConnell EE, Neal BH, Brunner MJ, Hamilton SB, Sullivan TM, et al. Comparative carcinogenicity in Sprague-Dawley rats of the polychlorinated biphenyl mixtures Aroclors 1016, 1242, 1254, and 1260. Toxicological Sciences. 1998;**41**(1):62-76

[20] Kashimoto T, Miyata H, Fukushima S, Kunita N, Ohi G, Tung TC. PCBs, PCQs and PCDFs in blood of yusho and yu-cheng patients. Environmental Health Perspectives. 1985;**59**:73-78

[21] Todaka T, Hirakawa H, Hori T, Tobiishi K, Iida T, Furue M. Concentrations of polychlorinated dibenzo-p-dioxins, polychlorinated dibenzofurans, and non-ortho and mono-ortho polychlorinated biphenyls in blood of Yusho patients. Chemosphere. 2007;**66**(10):1983-1989

[22] Hsu ST, Ma CI, Hsu SK, Wu SS, Hsu NH, Yeh CC, et al. Discovery and epidemiology of PCB poisoning in Taiwan: A four-year followup. Environmental Health Perspectives. 1985;**59**:5-10

[23] Guo YL, Ryan JJ, Lau BP, Yu ML, Hsu CC. Blood serum levels of PCBs and PCDFs in Yucheng women 14 years after exposure to a toxic rice oil. Archives of Environmental Contamination and Toxicology. 1997;**33**(1):104-108

[24] Onozuka D, Nakamura Y, Tsuji G, Furue M. Mortality in Yusho patients exposed to polychlorinated biphenyls and polychlorinated dibenzofurans: a 50-year retrospective cohort study. Environmental Health. 2020;**19**(1):119

[25] Li MC, Chen PC, Tsai PC, Furue M, Onozuka D, Hagihara A, et al. Mortality after exposure to polychlorinated biphenyls and polychlorinated dibenzofurans: A meta-analysis of two highly exposed cohorts. International Journal of Cancer. 2015;**137**(6):1427-1432

[26] Cogliano VJ. Assessing the cancer risk from environmental PCBs. Environmental Health Perspectives. 1998;**106**(6):317-323

[27] Abramowicz DA. Aerobic and anaerobic PCB biodegradation in the environment. Environmental health perspectives. 1995;**103**(Suppl 5):97-99

[28] Gilbert JM, Baduel C, Li Y, Reichelt-Brushett AJ, Butcher PA, McGrath SP, et al. Bioaccumulation of PCBs in liver tissue of dusky Carcharhinus obscurus, sandbar C. plumbeus and white Carcharodon carcharias sharks from south-eastern Australian waters. Marine Pollution Bulletin. 2015;**101**(2):908-913

[29] Cogliano VJ. PCBs: Cancer Dose-Response Assessment and Application to Environmental Mixtures. Washington, DC, EPA/600/P-96/001F., editor: U.S. Environmental Protection Agency OoRaD, National Center for Environmental Assessment, Washington Office; 1996

[30] Stecca L, Tait S, Corrado F, Esposito M, Mantovani A, La Rocca C. Development of an in vitro test battery model based on liver and colon cancer cell lines to discriminate PCB mixtures by transcription factors gene expression analysis. Toxicology In Vitro. 2016;**34**:204-211

[31] Shan Q, Li H, Chen N, Qu F, Guo J. Understanding the Multiple Effects of PCBs on Lipid Metabolism. Diabetes, Metabolic Syndrome and Obesity: Targets and Therapy. 2020;**13**:3691-3702

[32] Silberhorn EM, Glauert HP, Robertson LW. Carcinogenicity of polyhalogenated biphenyls: PCBs and PBBs. Critical Reviews in Toxicology. 1990;**20**(6):440-496

[33] Brunner MJ, Sullivan TM, Singer AW, Ryan MJ, Toft Ii JD, Menton RS, et al. An Assessment of the Chronic Toxicity and Oncogenicity of Aroclor-1016, Aroclor-1242, Aroclor-1254, and Aroclor-1260 Administered in Diet to Rats: Chronic Toxicity and Oncogenicity Report. Vol. 1. Columbus, OH: Battelle; 1997 Report No.: Battelle Study No.SC920192

[34] Norback DH, Weltman RH. Polychlorinated biphenyl induction of hepatocellular carcinoma in the Sprague-Dawley rat. Environmental Health Perspectives. 1985;**60**:97-105

[35] Rignall B, Grote K, Gavrilov A, Weimer M, Kopp-Schneider A, Krause E, et al. Biological and tumor-promoting effects of dioxin-like and non-dioxin-like polychlorinated biphenyls in mouse liver after single or combined treatment. Toxicological Sciences. 2013;**133**(1):29-41

[36] Haag-Grönlund M, Johansson N, Fransson-Steen R, Håkansson H, Scheu G, Wärngård L. Interactive effects of three structurally different polychlorinated biphenyls in a rat liver tumor promotion bioassay. Toxicology and Applied Pharmacology. 1998;**152**(1):153-165

[37] Stemm DN, Tharappel JC, Lehmler HJ, Srinivasan C, Morris JS, Spate VL, et al. Effect of dietary selenium on the promotion of hepatocarcinogenesis by 3,3′, 4,4′-tetrachlorobiphenyl and 2,2′, 4,4′, 5,5′-hexachlorobiphenyl. Experimental Biology and Medicine (Maywood, NJ). 2008;**233**(3):366-376

[38] Dean CE Jr, Benjamin SA, Chubb LS, Tessari JD, Keefe TJ. Nonadditive hepatic tumor promoting effects by a mixture of two structurally different polychlorinated biphenyls in female rat livers. Toxicological Sciences. 2002;**66**(1):54-61

[39] Zhang W, Sargis RM, Volden PA, Carmean CM, Sun XJ, Brady MJ. PCB 126 and other Dioxin-like PCBs specifically suppress hepatic PEPCK expression via the Aryl hydrocarbon receptor. PLoS One. 2012;**7**(5):e37103

[40] Glauert HP, Tharappel JC, Banerjee S, Chan NL, Kania-Korwel I, Lehmler HJ, et al. Inhibition of the promotion of hepatocarcinogenesis by 2,2′,4,4′,5,5′-hexachlorobiphenyl (PCB-153) by the deletion of the p50 subunit of NF-kappa B in mice. Toxicology and Applied Pharmacology. 2008;**232**(2):302-308

[41] Strathmann J, Schwarz M, Tharappel JC, Glauert HP, Spear BT, Robertson LW, et al. PCB 153, a non-dioxin-like tumor promoter, selects for beta-catenin (Catnb)-mutated mouse liver tumors. Toxicological Sciences. 2006;**93**(1):34-40

[42] Vincent F, de Boer J, Pfohl-Leszkowicz A, Cherrel Y, Galgani F. Two cases of ras mutation associated with liver hyperplasia in dragonets (Callionymus lyra) exposed to polychlorinated biphenyls and polycyclic aromatic hydrocarbons. Molecular Carcinogenesis. 1998;**21**(2):121-127

[43] Andersen O, Lindegaard P, Unger M, Nordberg GF. Effects of liver damage induced by polychlorinated biphenyls (PCB) on cadmium metabolism in mice. Environmental Research. 1985;**38**(2):213-224

[44] Whysner J, Wang CX. Hepatocellular iron accumulation and increased cell proliferation in polychlorinated biphenyl-exposed Sprague-Dawley rats and the development of hepatocarcinogenesis. Toxicological Sciences. 2001;**62**(1): 36-45

[45] Brown DP. Mortality of workers exposed to polychlorinated biphenyls--An update. Archives of Environmental Health. 1987;**42**(6):333-339

[46] Donato F, Moneda M, Portolani N, Rossini A, Molfino S, Ministrini S, et al. Polychlorinated biphenyls and risk of hepatocellular carcinoma in the population living in a highly polluted area in Italy. Scientific Reports. 2021;**11**(1):3064

[47] Zani C, Gelatti U, Donato F, Capelli M, Portolani N, Bergonzi R, et al. Polychlorinated biphenyls in serum, liver and adipose tissue of subjects with hepatocellular carcinoma living in a highly polluted area. Chemosphere. 2013;**91**(2):194-199

[48] Zhao G, Wang Z, Zhou H, Zhao Q. Burdens of PBBs, PBDEs, and PCBs in tissues of the cancer patients in the e-waste disassembly sites in Zhejiang, China. Science of the Total Environment. 2009;**407**(17):4831-4837

[49] Kaifie A, Schettgen T, Gube M, Ziegler P, Kraus T, Esser A. Functional and structural liver abnormalities in former PCB exposed workers – analyses from the HELPcB cohort. Journal of Toxicology and Environmental Health, Part A. 2019;**82**(1):52-61

[50] Clair HB, Pinkston CM, Rai SN, Pavuk M, Dutton ND, Brock GN, et al. Liver disease in a residential Cohort with elevated polychlorinated biphenyl exposures. Toxicological Sciences. 2018;**164**(1):39-49

[51] Loomis D, Browning SR, Schenck AP, Gregory E, Savitz DA. Cancer mortality among electric utility workers exposed to polychlorinated biphenyls. Occupational and Environmental Medicine. 1997;**54**(10):720-728

[52] Niehoff NM, Zabor EC, Satagopan J, Widell A, O'Brien TR, Zhang M, et al. Prediagnostic serum polychlorinated biphenyl concentrations and primary liver cancer: A case-control study nested within two prospective cohorts. Environmental Research. 2020;**187**:109690

[53] Delistraty D. Ecotoxicity and risk to human fish consumers of polychlorinated biphenyls in fish near the Hanford Site (USA). Science of the Total Environment. 2013;**445-446**:14-21

[54] Faust D, Vondráček J, Krčmář P, Smerdová L, Procházková J, Hrubá E, et al. AhR-mediated changes in global gene expression in rat liver progenitor cells. Archives of Toxicology. 2013;**87**(4): 681-698

[55] Borlak J, Jenke HS. Cross-talk between aryl hydrocarbon receptor and mitogen-activated protein kinase signaling pathway in liver cancer through c-raf transcriptional regulation. Molecular Cancer Research. 2008;**6**(8):1326-1336

[56] Buchmann A, Kunz W, Wolf CR, Oesch F, Robertson LW. Polychlorinated

biphenyls, classified as either phenobarbital- or 3-methylcholanthrene-type inducers of cytochrome P-450, are both hepatic tumor promoters in diethylnitrosamine-initiated rats. Cancer Letters. 1986;**32**(3):243-253

[57] Shen H, Robertson LW, Ludewig G. Regulatory effects of dioxin-like and non-dioxin-like PCBs and other AhR ligands on the antioxidant enzymes paraoxonase 1/2/3. Environmental Science and Pollution Research International. 2016;**23**(3):2108-2118

[58] Oakley GG, Devanaboyina U, Robertson LW, Gupta RC. Oxidative DNA damage induced by activation of polychlorinated biphenyls (PCBs): Implications for PCB-induced oxidative stress in breast cancer. Chemical Research in Toxicology. 1996;**9**(8):1285-1292

[59] Birnbaum LS. Endocrine effects of prenatal exposure to PCBs, dioxins, and other xenobiotics: Implications for policy and future research. Environmental Health Perspectives. 1994;**102**(8):676-679

[60] Andersson PL, Blom A, Johannisson A, Pesonen M, Tysklind M, Berg AH, et al. Assessment of PCBs and hydroxylated PCBs as potential xenoestrogens: In vitro studies based on MCF-7 cell proliferation and induction of vitellogenin in primary culture of rainbow trout hepatocytes. Archives of Environmental Contamination and Toxicology. 1999;**37**(2):145-150

[61] Ho PW, Garner CE, Ho JW, Leung KC, Chu AC, Kwok KH, et al. Estrogenic phenol and catechol metabolites of PCBs modulate catechol-O-methyltransferase expression via the estrogen receptor: Potential contribution to cancer risk. Current Drug Metabolism. 2008;**9**(4):304-309

[62] Deng P, Barney J, Petriello MC, Morris AJ, Wahlang B, Hennig B. Hepatic metabolomics reveals that liver injury increases PCB 126-induced oxidative stress and metabolic dysfunction. Chemosphere. 2019;**217**:140-149

[63] Yuan J, Lu WQ, Zou YL, Wei W, Zhang C, Xie H, et al. Influence of aroclor 1254 on benzo(a)pyrene-induced DNA breakage, oxidative DNA damage, and cytochrome P4501A activity in human hepatoma cell line. Environmental Toxicology. 2009;**24**(4): 327-333

[64] Al-Anati L, Högberg J, Stenius U. Non-dioxin-like-PCBs phosphorylate Mdm2 at Ser166 and attenuate the p53 response in HepG2 cells. Chemico-Biological Interactions. 2009;**182**(2-3): 191-198

[65] Chen Y, Wu Y, Xiao W, Jia H, Glatt H, Shi M, et al. Human CYP1B1-dependent genotoxicity of dioxin-like polychlorinated biphenyls in mammalian cells. Toxicology. 2020;**429**:152329

[66] Hu K, Yu H, Li Z, Jin G, Jia H, Song M, et al. Human CYP2E1-activated mutagenicity of dioxin-like PCBs 105 and 118-Experimental data consistent with molecular docking results. Toxicology. 2020;**437**:152438

[67] De Haan LH, Simons JW, Bos AT, Aarts JM, Denison MS, Brouwer A. Inhibition of intercellular communication by 2,3,7,8-tetrachlorodibenzo-p-dioxin and dioxin-like PCBs in mouse hepatoma cells (Hepa1c1c7): Involvement of the Ah receptor. Toxicology and Applied Pharmacology. 1994;**129**(2):283-293

[68] VanEtten SL, Bonner MR, Ren X, Birnbaum LS, Kostyniak PJ, Wang J, et al. Telomeres as targets for the toxicity of 2,3,7,8-tetrachlorodibenzo-p-dioxin (TCDD) and polychlorinated biphenyls (PCBs) in rats. Toxicology and Applied Pharmacology. 2020;**408**:115264

[69] de Conti A, Beland FA, Pogribny IP. The role of epigenomic alterations in furan-induced hepatobiliary pathologies. Food and Chemical Toxicology. 2017;**109**:677-682

[70] Lu D, Sullivan MM, Phillips MB, Peterson LA. Degraded protein adducts of cis-2-butene-1, 4-dial are urinary and hepatocyte metabolites of furan. Chemical Research in Toxicology. 2009;**22**(6):997-1007

[71] Grill AE, Schmitt T, Gates LA, Lu D, Bandyopadhyay D, Yuan J-M, et al. Abundant rodent furan-derived urinary metabolites are associated with tobacco smoke exposure in humans. 2015;**28**(7):1508-1516

[72] Knutsen HK, Alexander J, Barregård L, Bignami M, Brüschweiler B, Ceccatelli S, et al. Risks for public health related to the presence of furan and methylfurans in food. EFSA journal European Food Safety Authority 2017;**15**(10):e05005

[73] Mariotti MS, Granby K, Rozowski J, Pedreschi FJF. Furan: a critical heat induced dietary contaminant. Food & Function. 2013;**4**(7):1001-1015

[74] LAJ P. Electrophilic intermediates produced by bioactivation of furan. Drug Metabolism Reviews. 2006;**38**(4):615-626

[75] Phillips MB, Sullivan MM, Villalta PW, Peterson LA. Covalent modification of cytochrome c by reactive metabolites of furan. Chemical Research in Toxicology. 2014;**27**(1):129-135

[76] Nunes J, Martins IL, Charneira C, Pogribny IP, de Conti A, Beland FA, et al. New insights into the molecular mechanisms of chemical carcinogenesis: In vivo adduction of histone H2B by a reactive metabolite of the chemical carcinogen furan. 2016;**264**:106-113

[77] Jackson AF, Williams A, Recio L, Waters MD, Lambert IB, Yauk CLJT, et al. Case study on the utility of hepatic global gene expression profiling in the risk assessment of the carcinogen furan. 2014;**274**(1):63-77

[78] Gill S, Kavanagh M, Barker M, Weld M, Vavasour E, Hou Y, et al. Subchronic oral toxicity study of furan in B6C3F1 Mice. Toxicol Pathol. 2011;**39**(5):787-794

[79] Ramm S, Limbeck E, Mally A. Functional and cellular consequences of covalent target protein modification by furan in rat liver. Toxicology. 2016;**361**:49-61

[80] Moro S, Chipman JK, Antczak P, Turan N, Dekant W, Falciani F, et al. Identification and pathway mapping of furan target proteins reveal mitochondrial energy production and redox regulation as critical targets of furan toxicity. 2012;**126**(2):336-352

[81] Selmanoğlu G, Karacaoğlu E, Kılıç A, Koçkaya EA, Akay MT. Toxicity of food contaminant furan on liver and kidney of growing male rats. Environmental Toxicology. 2012;**27**(10):613-622

[82] Terrell AN, Huynh M, Grill AE, Kovi RC, O'Sullivan MG, Guttenplan JB, et al. Mutagenicity of furan in female Big Blue B6C3F1 mice. 2014;**770**:46-54

[83] Ding W, Petibone DM, Latendresse JR, Pearce MG, Muskhelishvili L, White GA, et al. In vivo genotoxicity of furan in F344 rats at cancer bioassay doses. 2012;**261**(2):164-171

[84] Takasu S, Ishii Y, Kijima A, Ogawa K, Nakane S, Umemura T. Furan induced characteristic glutathione S-transferase placental form-positive Foci in terms of cell kinetics and gene expression. Toxicologic Pathology. 2020;**48**(6):756-765

[85] Hibi D, Yokoo Y, Suzuki Y, Ishii Y, Jin M, Kijima A, et al. Lack of genotoxic

mechanisms in early-stage furan-induced hepatocellular tumorigenesis in gpt delta rats. 2017;**37**(2):142-149

[86] Hickling K, Hitchcock J, Chipman J, Hammond T, Evans JJ. Induction and progression of cholangiofibrosis in rat liver injured by oral administration of furan. Toxicologic Pathology. 2010;**38**(2):213-229

[87] Hickling KC, Hitchcock JM, Oreffo V, Mally A, Hammond TG, Evans JG, et al. Evidence of oxidative stress and associated DNA damage, increased proliferative drive, and altered gene expression in rat liver produced by the cholangiocarcinogenic agent furan. Toxicol Pathol. 2010;**38**(2):230-243

[88] Tryndyak V, de Conti A, Doerge DR, Olson GR, Beland FA, Pogribny IP. Furan-induced transcriptomic and gene-specific DNA methylation changes in the livers of Fischer 344 rats in a 2-year carcinogenicity study. Archives of Toxicology. 2017;**91**(3):1233-1243

[89] Durling JKL, Svensson K, Abramsson-Zetterberg L. Furan is not genotoxic in the micronucleus assay in vivo or in vitro. Toxicology Letters. 2007;**169**(1):43-50

[90] McDaniel LP, Ding W, Dobrovolsky VN, Shaddock JG Jr, Mittelstaedt RA, Doerge DR, et al. Genotoxicity of furan in Big Blue rats. 2012;**742**(1-2):72-78

[91] Chen T, Williams TD, Mally A, Hamberger C, Mirbahai L, Hickling K, et al. Gene expression and epigenetic changes by furan in rat liver. 2012;**292**(2-3):63-70

[92] Conti A, Kobets T, Escudero-Lourdes C, Montgomery B, Tryndyak V, Beland FA, et al. Dose-and time-dependent epigenetic changes in the livers of Fisher 344 rats exposed to furan. 2014;**139**(2):371-380

[93] Recio L, Phillips SL, Maynor T, Waters M, Jackson AF, Yauk CL. Differential expression of long noncoding RNAs in the livers of female B6C3F1 mice exposed to the carcinogen furan. Toxicological Sciences. 2013;**135**(2):369-379

[94] Wong MH, Armour M-A, Naidu R, Man M. Persistent toxic substances: Sources, fates and effects. Reviews on Environmental Health. 2012;**27**(4):207-213

[95] English JSC, Dawe RS, Ferguson JJ. Environmental effects and skin disease. British Medical Bulletin. 2003;**68**(1): 129-142

[96] Ju Q, Zouboulis CJB. Effects of 2, 3, 7, 8-tetrachlorodibenzo-p-dioxin on the skin. 2013;**2, 113** 2167-0331.1000113

[97] Pelclová D, Urban P, Preiss J, Lukáš E, Fenclová Z, Navrátil T, et al. Adverse health effects in humans exposed to 2, 3, 7, 8-tetrachlorodibenzo-p-dioxin (TCDD). 2006;**21**(2):119-138

[98] Çiftçi O. The Investigation of effect mechanism, chemical structure and toxicokinetics properties of dioxins compounds. Annals of Medical reaserch. 2010;**17**(4):413-422

[99] Bentli R, Ciftci O, Cetin A, Otlu AJT. Montelukast, an antiinflamattory agent, can prevent toxic effects of 2, 3, 7, 8-tetrachlorodibenzo-p-dioxin in terms of oxidative stress and histological alterations in liver and serum cytokine levels. Toxicology and Industrial Health. 2013

[100] Türkez H, Geyikoğlu F, Yousef MI, Toğar B, Vançelik SJT. Propolis alleviates 2, 3, 7, 8-Tetrachlorodibenzo-p-dioxin-induced histological changes, oxidative stress and DNA damage in rat liver. Toxicology and Industrial Health. 2013;**29**(8):677-685

[101] Mejia-Garcia A, Sanchez-Ocampo EM, Galindo-Gomez S,

Shibayama M, Reyes-Hernandez O, Guzman-Leon S, et al. 2, 3, 7, 8-Tetrachlorodibenzo-p-dioxin enhances CCl4-induced hepatotoxicity in an aryl hydrocarbon receptor-dependent manner. 2013;**43**(2):161-168

[102] Lee JH, Wada T, Febbraio M, He J, Matsubara T, Lee MJ, et al. A novel role for the dioxin receptor in fatty acid metabolism and hepatic steatosis. 2010;**139**(2):653-663

[103] Gonzalez JF, Fernandez-Salguero P. The aryl hydrocarbon receptor: Studies using the AHR-null mice. Drug Metabolism and Disposition. 1998;**26**(12):1194-1198

[104] Elizondo G, Fernandez-Salguero P, Sheikh MS, Kim G-Y, Fornace AJ, Lee KS, et al. Altered cell cycle control at the G2/M phases in aryl hydrocarbon receptor-null embryo fibroblast. 2000;**57**(5):1056-1063

[105] Sato S, Shirakawa H, Tomita S, Ohsaki Y, Haketa K, Tooi O, et al. Low-dose dioxins alter gene expression related to cholesterol biosynthesis, lipogenesis, and glucose metabolism through the aryl hydrocarbon receptor-mediated pathway in mouse liver. 2008;**229**(1):10-19

[106] Reyes-Hernandez O, Mejia-Garcia A, Sánchez-Ocampo E, Cabanas-Cortes M, Ramirez P, Chávez-González L, et al. Ube2l3 gene expression is modulated by activation of the aryl hydrocarbon receptor: Implications for p53 ubiquitination. 2010;**80**(6):932-940

[107] Walisser JA, Glover E, Pande K, Liss AL, Bradfield CA. Aryl hydrocarbon receptor-dependent liver development and hepatotoxicity are mediated by different cell types. Proceedings of the National Academy of Sciences of the United States of America. 2005;**102**(49):17858-17863

[108] Boutros PC, Bielefeld KA, Pohjanvirta R, Harper PA. Dioxin-dependent and dioxin-independent gene batteries: Comparison of liver and kidney in AHR-null mice. Toxicological Sciences. 2009;**112**(1):245-256

[109] Kennedy GD, Nukaya M, Moran SM, Glover E, Weinberg S, Balbo S, et al. Liver tumor promotion by 2, 3, 7, 8-tetrachlorodibenzo-p-dioxin is dependent on the aryl hydrocarbon receptor and TNF/IL-1 receptors. 2014;**140**(1):135-143

[110] Yamaguchi M, Hankinson O. 2, 3, 7, 8-Tetrachlorodibenzo-p-dioxin suppresses the growth of human liver cancer HepG2 cells in vitro: Involvement of cell signaling factors. International Journal of Oncology. 2018;**53**(4): 1657-1666

[111] Türkez H, Geyikoglu F, Yousef MI. Ameliorative effect of docosahexaenoic acid on 2, 3, 7, 8-tetrachlorodibenzo-p-dioxin-induced histological changes, oxidative stress, and DNA damage in rat liver. Toxicology and Industrial Health. 2012;**28**(8):687-696

[112] Alsharif NZ, Hassoun EA. Protective effects of vitamin A and vitamin E succinate against 2, 3, 7, 8-tetrachlorodibenzo-p-dioxin (TCDD)-induced body wasting, hepatomegaly, thymic atrophy, production of reactive oxygen species and DNA damage in C57BL/6J mice. Basic & Clinical Pharmacology & Toxicology. 2004;**95**(3):131-138

[113] Hassoun EA, Vodhanel J, Abushaban A. The modulatory effects of ellagic acid and vitamin E succinate on TCDD-induced oxidative stress in different brain regions of rats after subchronic exposure. Journal of Biochemical and Molecular Toxicology. 2004;**18**(4):196-203

[114] Ahd K, Dhibi S, Akermi S, Bouzenna H, Samout N, Elfeki A, et al. Protective effect of ginger (Zingiber

officinale) against PCB-induced acute hepatotoxicity in male rats. 2019;**9**(50):29120-29130

[115] Ciftci O, Ozdemir I. Protective effects of quercetin and chrysin against 2, 3, 7, 8-tetrachlorodibenzo-p-dioxin (TCDD) induced oxidative stress, body wasting and altered cytokine productions in rats. 2011;**33**(3):504-508

[116] Kim H-S, Park S-Y, Yoo K-Y, Lee SK, Jung W-W. Induction of heat shock proteins and antioxidant enzymes in 2, 3, 7, 8-TCDD-induced hepatotoxicity in rats. The Korean Journal of Physiology & Pharmacology. 2012;**16**(6):469-476

[117] Czepiel J, Biesiada G, Gajda M, Szczepański W, Szypuła K, Dąbrowski Z, et al. The effect of TCDD dioxin on the rat liver in biochemical and histological assessment. 2009;**58**(1-2):85-90

[118] Viluksela M, Pohjanvirta R. Multigenerational and transgenerational effects of dioxins. International Journal of Molecular Sciences. 2019;**20**(12):2947

[119] Boutros PC, Yan R, Moffat ID, Pohjanvirta R, Okey AB. Transcriptomic responses to 2, 3, 7, 8-tetrachlorodibenzo-p-dioxin (TCDD) in liver: comparison of rat and mouse. BMC Genomics. 2008;**9**(1):1-17

[120] Boverhof DR, Burgoon LD, Tashiro C, Sharratt B, Chittim B, Harkema JR, et al. Comparative toxicogenomic analysis of the hepatotoxic effects of TCDD in Sprague Dawley rats and C57BL/6 mice. 2006;**94**(2):398-416

[121] Zhang X, Li C, Gong Z. Development of a convenient in vivo hepatotoxin assay using a transgenic zebrafish line with liver-specific DsRed expression. 2014;**9**(3):e91874

[122] Yang Q, Salim L, Yan C, Gong Z. Rapid Analysis of Effects of Environmental Toxicants on Tumorigenesis and Inflammation Using a Transgenic Zebrafish Model for Liver Cancer. Marine Biotechnology (New York, NY). 2019;**21**(3):396-405

Chapter 4

Persistent Organic Pollutants in the Bizerte Lagoon Ecosystem: Occurrence, Distribution, and Ecotoxicological Assessment Using Marine Organism

Ahmed Nasri, Takoua Mhadhbi, Mohamed Allouche, Amel Hannachi, Hary Demey, Patricia Aïssa, Hamouda Beyrem and Ezzeddine Mahmoudi

Abstract

Marine ecosystem represents an ecologically and economically important water bodies for human and animal living. Their increasing pollution by persistent organic pollutants has represented a major environmental alarm during the last years. In the current study, we examined the occurrence, local distribution and ecotoxicological menace of organic pollutants, comprising brominated flame retardants (BFR), polychlorinated biphenyls (PCBs), polycyclic aromatic hydrocarbons (PAHs), and organochlorine pesticides (OCPs) in different matrices from the Northern Tunisian Coastal Ecosystem (Bizerte lagoon). The pollutant existence in this biome is related with a negative impact on the biocenosis health. Many approach including (i) chemical analyses; (ii) taxonomic structure and ecological indices analyses; (iii) and biochemical experimental studies, were investigated to determine the ecosystem quality and the contaminant effects. Our chapter introduces the baseline information on the organic contaminations extent and toxicological impact, as well as, it contribute to evaluate the ecological quality of this marine coastal ecosystem.

Keywords: Persistent Organic Pollutants (POPs), Mediterranean coastal lagoon, Occurrence, Distribution, Ecotoxicological evaluation

1. Introduction

The expansion of the anthropoid industrial activities has led to the pollution of many ecosystems. The daily release of agricultural, industrial and commercial chemicals into the aquatic ecosystem has induced various toxic effects on marine organisms [1]. Among Mediterranean coastal environments, many lagoons are exposed to human-induced pressures. The Bizerte lagoon is one of them, located near an industrial and agricultural area of Northern Tunisia, known as a receptor of numerous industrial wastes, chemical fertilizers and pesticides via runoff and soil erosion, conducting to a lessening in fish and bivalves' production [2].

These organisms accumulate chemicals directly from contaminated water and indirectly via the food chain. In addition, these are used as sentinels' species because they have the capacity to bio-concentrate toxic compounds. Their main advantage is the rapid response to low concentrations of aquatic contaminant [3].

Biomonitoring of the aquatic environment has based both on the quantification of pollutant concentrations in bioindicator organisms and on the biological analyses (ecological indices and biomarkers). The chemical analyses were performed after pollutants extraction from different matrices (water, sediment, and marine organism), using instruments by Gas chromatography separation connected or not to a mass spectrometer (GC–MS) [4–15]. Nevertheless, the biological analyses were performed both firstly; by nematodes taxonomic structure and ecological indices determination to assess the chemicals toxicity, secondly, by biomarkers assessment who are known as tools for contamination impacts evaluation in the marine environment. Some of these have been incorporated in ecological monitoring programs [16].

The biomarkers have been classified in different types, such as biomarkers exposure and response, or as general and specific biomarkers [17]. Antioxidant enzymes such as catalase (CAT), reduced glutathione (GSH) and Glutathione-S-transferase (GST) have been used as biomarkers reflecting the exposure and toxicity of contaminants [18]. The Acetylcholinesterase (AChE) activity has been used as a biomarker of exposure to several pesticides (organophosphate and carbamate insecticides) in aquatic environments [19]. The high concentrations of Malondialdehyde (MDA) have been reflected the lipid peroxidation expression that indicates cell damage [20].

This aim of the present chapter was to describe the previous studies focused on the determination of pollutant concentrations in bio-indicator organisms and to discuss the results of biological assessment of their toxicities in order to provide clearer and more informative data concerning the Bizerte lagoon coastal ecosystem state.

2. Persistent organic pollutants (POP)

2.1 Definition

Persistent organic pollutants (POPs) are organic chemical compounds having physical and chemical properties that result in their widespread dissemination, persistence and accumulation. They are known by their characteristics: persistence, bioaccumulation, transport, and toxicity effects. POPs have a lifespan that can reach many years, because of their resistance to chemical and biological degradation. They are lipophilic and tend to disperse widely across ecosystems; they accumulate to high concentrations in the wildlife tissues biomagnifying up the food chain. They are semi-volatile, allowing for long-range transport in air or absorbed to particulate matter in environmental matrices such as air or water.

These chemicals are organic substances of natural or anthropogenic origin, once released into the environment; they remain intact for exceptionally long periods of time [21]. The major intentional man-made sources of POPs are industrial sources like incinerating plants, power stations, agricultural sprays, heating stations and evaporation from water surfaces, soil, or from the landfills.

3. Sampling location

3.1 Bizerte lagoon

The Bizerte Lagoon is located in the Tunisia northern littoral (Mediterranean Sea) (latitude, 37°80′–37°14′ N; longitude, 9°46′–9°56′ E). It spread 128 km^2 of

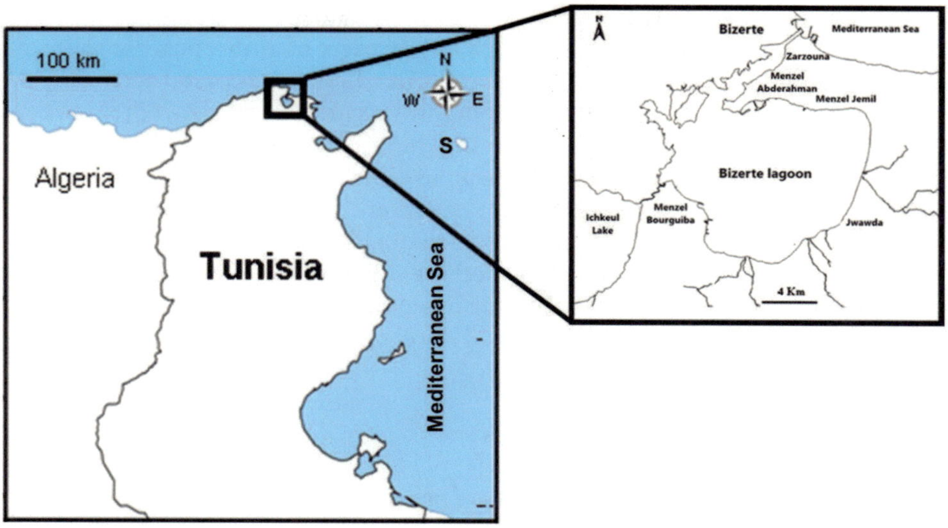

Figure 1.
Location of the Bizerte lagoon (northern Tunisia).

area and 7 m of mean depth. In the north it communicates by a 7-km-long canal with the Mediterranean Sea, nonetheless, in the South with the Ichkeul Lake by the River of Tinja. This lagoon is under tributaries of many rivers such as Gueniche, Ben Hassine, Mrezig, Garek, and Tinja has a high importance in terms of seafood products exports; it show an economically significant body of water due to a variety of peach and aquaculture activities (oysters and mussels farming).

Due the population growth and technological development around the lagoon, this ecosystem is compelling to the influence of divers' physical factors that highly fluctuate during the year. It is severely affected by human activities on its shores, including the four harbors presence, urban development (bounded by the cities of Bizerte, Zarzouna, Menzel Jemil, Menzel Abderrahmane, Menzel Bourguiba and Tinja), and receives urban effluents, agricultural inputs and industrial discharges from many industrial units related to various fields of activity such as cement, steel plant, a metal factory, and a refinery (**Figure 1**) [22].

In the lagoon of Bizerte, the water column is under the influence of the winter contributions of rains and fresh water in provenance of the wadis, particularly that of Ichkeul [23]. On the other hand, the water column is under the influence of the winter contributions of rains and fresh water in provenance of the wadis, particularly that of Ichkeul [23]. On the other hand, the water exchange between the Mediterranean and the Ichkeul make the temperature and salinity of the lagoon vary between (10–29°C) and (32.5–38.5‰), respectively.

4. POPs in the Bizerte lagoon

Due to their persistent nature and ability to undergo long-range transport, POPs are ubiquitous in the environment. They are found in soil, lake and river sediment, benthos, water columns and the atmosphere. They enter and accumulate in the food chain through various pathways, such as deposition onto crops, soil ingestion by grazing animals, and bioaccumulation up through trophic levels. At concentrations typically found in food, the adverse health effects caused by POPs are almost entirely chronic (including cancer), disruption of the endocrine system, neurotoxicity, and development damage.

The occurrence of POPs is reported major in the aquatic environment pollution of which released into water bodies from a large variety of anthropogenic sources such as agricultural and municipal waste effluents, industrial coastal activities, atmospheric deposition, maritime transport and accidental spill [24, 25]. Many studies have been carried out for monitoring the organic pollutants and their impact on different aquatic ecosystems around the Bizerte lagoon. The most identified POP chemicals in Bizerte lagoon ecosystems are belonging the families of brominated (and chlorinated) aromatics, including brominated flame retardants (BFR) such as polybrominated diphenyl ethers (PBDEs) and their methoxylated analogues (MeO-PBDEs) [4–9]; polychlorinated biphenyls (PCBs) [10–12]; polycyclic aromatic hydrocarbons PAHs [13, 14]; and organochlorine pesticides (OCPs) [11, 12, 15].

4.1 Brominated flame retardants (BFRs)

Brominated flame retardants (BFRs) are brominated organic compounds used as additive flame retardants in plastics, paints, textiles, electronics, and vehicles. PBDEs and analogs (such as the methoxylated PBDEs (MeO-PBDEs)), have been found in many environmental matrices such as, aquatic organism, water, sludge, and sediment [26], and in fish and marine mammals [27], respectively. Contamination by BFRs is of environmental concern due to their persistence, potential for endocrine disruption and bioaccumulation, and long-range transport. It has been reported their adverse effect, such as morphological, immunological, and behavioral modifications, and enzyme induction [28, 29].

The levels of the BFRs in two species like in sea bass (*Dicentrarchus labrax*) and mullet (*Mugil cephalus*) collected from the Bizerte Lagoon were examined. The PBDE mean concentrations in fish were 45.3 and 96.2 ng/g lw respectively in mullet and sea bass. MeO PBDE concentrations in mullet and sea bass were 6.46–286 ng/g lw and 49.4–798 ng/g lw, respectively [4]. In other study, polybrominated diphenyl-ethers (PBDEs) and methoxylated polybrominated diphenyl ethers (MeO-PBDEs) were determined in *solea solea* muscle. Mean levels of these compounds were 279 ng/g lw in sole [5]. In addition, sediment and urchins were used for the levels analysis of halogenated flame retardants (HFRs) and the methoxylated-PBDEs. The sediment concentrations were found between nd – 51.8 ng/g dry weight (dw) for the HFRs, and not detected for rest compounds. However, the levels of polybrominated diphenyl ethers (PBDEs), halogenated norbornenes (HN) and methoxylated PBDEs (MeO-PBDEs) were 3.67 to 56.9 ng/g lipid weight (lw), 4.52 to 116 ng/g lw and nd - 364 ng/g lw, respectively, in sea urchins [8].

Ameur et al. determined the concentrations of HFRs and MeO-PBDEs in *Hexaplex trunculus* collected in the same ecosystem and found that the mean tissue levels vary between 187 and 264 ng/g lw [6]. Recently, Mekni et al. measured the presence of the Brominated flame-retardants in sediment and fish eel (*Anguilla Anguilla*) samples [9]. Sediment HFR (halogenated flame-retardants) levels were 3.30–28.5 ng/g dry weight (dw), while OPFR (organophosphate flame retardants) levels were 9.77–164 ng/g dw [9]. However, the levels were 4.72–151 ng/g lipid weight (lw) and 19.7–2154 ng/g lw were estimated in fish, respectively.

4.2 Polychlorinated biphenyls (PCBs)

PCBs are produced and used in a wide range of industrial applications, such as inks, coatings, electrical transformers, and paints. Despite their prohibitions, they can also be found in various compartments such as sediments [4, 10, 30], in biota [31, 32], in human blood, breast milk and serum [33, 34].

Concentrations of polychlorinated biphenyls (PCBs) were determined in sediments at range between 0.89 and 6.63 ng/g dw [10], and in fish species collected from Bizerte Lagoon were found 164 to 336 ng/g lw and from 282 to 642 ng/g lw, respectively in mullet (*Mugil cephalus*) and sea bass (*Dicentrarchus labrax*). PCB-118, PCB-138, PCB-153 and PCB-180 were dominant contaminants in the studied fish species, accounting respectively for 9.00%, 14.0%, 28.5% and 23.6% of total PCBs [11]. Polychlorinated biphenyls (PCBs) were measured in fish (*solea solea*) muscle and the mean levels were 1417 and 315 ng/g lipid weight (lw) [5]. Recently, the concentration of polychlorinated biphenyls (PCBs) identified in water samples were ranged between 3 and 10.4 ng L^{-1}. PCB-28 (0.2–1.4 ng L^{-1}) and PCB-52 (0.8–3.5 ng L^{-1}) were the predominant PCB congeners [35].

4.3 Polycyclic aromatic hydrocarbons (PAHs)

Polycyclic aromatic hydrocarbons (PAHs) are an important class of persistent organic pollutants derived from anthropogenic or natural sources. They are formed as a consequence of incomplete combustion of organic matter; they are also found in the crude oil components and derived products. Natural sources of PAHs include forest-fires and post-depositional transformation of biogenic precursors [36]. PAHs reach the marine environment via such sources as effluent discharges, urban run-off, atmospheric transport, and agricultural runoff. Because of their low water solubility and their hydrophobicity, PAHs in the aquatic environment rapidly become associated with inorganic and organic suspended particles [37, 38] and subsequent deposition in sediment. PAHs accumulate in aquatic sediments after adsorption to particles due to their high hydrophobicity and low solubility [38, 39].

Total PAHs identified in sediments from 10 stations around the Bizerte lagoon (Menzel Abderahmen, Menzel Jemil, Oued Guenniche, Oued Garek, Oued Ben Hasssin, Mnzel Bourguiba...) were ranged between 83.3–447 (ng/g dry wt) [13] with a mean value of 218 ng/g dry wt. Only anthracene, acenaphthene, and phenanthrene are found in all stations. Recently, Barhoumi et al. have been showed that form many station (n = 18), the total concentration varied between 16.9 to 394.1 (ng/g dry wt) with a mean concentration of 85.5 ng/g dry wt [14]. The maximum levels of PAHs were found along the channel and were ranged between 160.2–394.1 (ng/g dw), followed by station at the nearness of the mouth of the Tinja River which is affected by agricultural inputs (102.8 ng/g dw). The same researchers have showed the presence of PAHs levels in two different species, mussels (*Mytilus galloprovincialis*) (107.4 to 430.7 ng/g dw) and fish (*Anguilla anguilla*) (114.5–133.7 ng/g dw). In addition, Naphthalene was the most important hydrocarbon identified in mussels (31.5–272.6 ng/g dw) and fish (57.9–68.6 ng/g dw) [40]. More recently, 16 PAHs were measured in 40 surface sediment samples from the Bizerte lagoon and the high concentrations of total PAHs were 122–19600 ng/g [15].

4.4 Organochlorine pesticides (OCPs)

The organochlorine pesticides (OCPs) represent one of the most known persistent organic pollutants (POPs), which have caused great concern all over the world as a result of their persistence for many years, high lipophilicity, long range transportation, chronic and acute toxicities and bioaccumulation [41]. The OCPs have been produced and used for agricultural and industrial purposes for a long time and on a large scale. Organochlorine pesticides had been used throughout the world thanks to exceptional insecticidal and fungicidal properties [42].

The OCPs concentrations were measured in fish such as *Mugil cephalus* and *Dicentrarchus labrax* and the mean levels were found from 52.9 to 157 ng/g lw and from 158 to 265 ng/g lw, respectively [11]. In other study, Barhoumi et al. have found that the total sediments concentrations of OCPs from 1.1 to 14.0 ng/g dw (average value, 3.3 ng/g dw) [12]. Among the OCPs, the concentrations range of DDTs (dichlorodiphenyltrichloroethane) in addition with its metabolites and HCB (hexachlorobenzene) were 0.3–11.5 ng/g dw (1.9 ng/g dw) and 0.6–2.5 ng/g dw (1.4 ng/g dw), respectively. Lately, sediment has examined for study the vertical distribution of organochlorine pesticides (OCPs). The OCPs concentrations ranged from 26.98 ± 0.04 ng/g found at 3 cm depth and 10.23 ± 0.02 ng/g at 6 cm depth in site SC1 (located in the mouth of the channel connecting the lagoon to the Mediterranean Sea). However, in station SC2 (located in the metallurgy of Menzel bourguiba), the concentrations are of the order of 11.77 ± 0.11 ng/g (9 cm deep) and 1.47 ± 0.02 ng/g (20 cm deep) [43].

5. POPs impact on marine organism

The occurrence of chemicals compounds in aquatic environment has led researchers to examine biological effects on aquatic organisms. Persistent organic pollutants (POPs) are toxic chemicals that adversely affect organism health and the environment. Many effects are associated to POPs exposure such as immune alteration, reproductive disorder, endocrine disruption, and neurological disorders. In the Bizerte lagoon, the use of marine organisms as bio-indicators of contaminant loads (and as models of laboratory toxicity studies), has enabled the gathering of information on the state of the ecosystem under consideration.

Laboratory research has documented many effects of POPs in a wide range of aquatic organism. Toxicity of these compounds are associated to various biochemical and population effect in benthic nematodes, marine phytoplankton, and bivalves. Among these studies, Louati et al. studied the response of microbial communities following sediment enrichment in anthracene as well as the monitoring of their biomass, activity and composition following the bioremediation [44]. A significant reduction of bacterial abundance, a strongly oxygen consumption inhibition, as well as, a microbial structure modification in comparison with the control were registered. In other study, marine nematodes communities were subjected to a 100 ppm polycyclic aromatic hydrocarbons mixture of phenanthrene, fluoranthene, and pyrene during 30 days. Abundance, diversity, and taxonomic structure were changed. *Spirinia parasitifera* became the resistant species to PAHs contamination while *Oncholaimus campylocercoides* and *Neochromadora peocilosoma* were strongly inhibited [45].

Othman et al. studied the toxicity of PAHs mixtures on natural phytoplankton communities [46]. Results show that PAHs decreased the photosynthetic potential with a dramatic change in phytoplankton community composition (size classes and chlorophyll a) were strongly affected. The diatom *Entomoneis paludosa* appeared favored under PAH exposure as evidenced by increase in cell density, whereas autotrophic flagellates and dinophytes were strongly reduced. In other study, the anthracene toxicity was studied on the benthic bivalves *Ruditapes decussatus.* A change of the siphon movement and decreasing filtration rate were recorded. In addition, the oxidative stress status of the gills was affected with the modification of proteins [47]. Exposure of the pelagic organism (*Mytilus galloprovincialis*) to anthracene induced the enzymatic activities such as, acetylcholinesterase (AChE), glutathione (GSH), and malondialdehyde (MDA) in digestive gland. A reduction of the filtration rates and an increase in lipid peroxidation in digestive gland were registered. Finally, the GSH content as well as AChE activity have been reduced in digestive gland [48]. Study examined the Benzo[a] Pyrene (B[a]P) toxicity in

Mytilus galloprovincialis and *Ruditapes decussatus* showed an increase of the total oxyradical scavenging capacity (TOSC) in the digestive gland after exposure to 100 and 300 μg/L concentrations of B[a]P. The superoxide dismutase (SOD), glutathione S-transferase (GST), and catalase (CAT) activities in gills and digestive gland were significantly induced. A significant increase of the GST activity and a decrease of AChE activity in digestive glands and gills were recorded in the two species. Interestingly, an increase of MDA level in the gills and digestive gland only in *Ruditapes decussatus* species [49].

Recently, Nasri et al. investigated the meiobenthic response to the polybrominated diphenyl ether (BDE-47) exposure in laboratory during one month [50]. BDE-47 caused a decrease in the bacterial abundance, and the nematodes taxonomic diversity as well as a change of all functional traits abundance, especially, the feeding group, amphid shape, and adult length were the most affected. In other study, Nasri et al. determined the taxonomic and trophic response of marine nematodes with the same concentrations of BDE-47 [51]. Species abundance and all univariate indices were significantly affected. BDE-47 treatment caused the microvores group represented by two species of *Terschellingia* to be replaced by the more resistant trophic groups such as epigrowth feeders (*Paracomesoma dubium*) and facultative predators (*Metoncholaimus pristiurus*).

6. Conclusions

The uses of chemical and biological analyses are promising tools for determining the quality of the Bizerte lagoon ecosystem and to survey the effects of the contaminants from anthropological activities. As a result, several studies have shown the effectiveness of the use in addition of sediments, associated organisms, such as marine bivalves, to monitor pollution [52]. They are specific indicators of different matrices in the ecosystem in relation to their position, and they show different rates of biotransformation and bioaccumulation compared to xenobiotics [53].

Conflict of interest

The authors declare no conflict of interest.

Author details

Ahmed Nasri[1*], Takoua Mhadhbi[1], Mohamed Allouche[1], Amel Hannachi[1], Hary Demey[2], Patricia Aïssa[2], Hamouda Beyrem[1] and Ezzeddine Mahmoudi[1]

1 Faculty of Sciences of Bizerte (FSB), Laboratory of Environment Biomonitoring, University of Carthage, Zarzouna, Bizerta, Tunisia

2 Department of Chemical Engineering, ETSEIB, Universitat Politècnica de Catalunya, Barcelona, Spain

*Address all correspondence to: a7mednas@gmail.com; ahmed.nasri@fsb.u-carthage.tn

References

[1] Nasri A et al. Ethinylestradiol (EE2) residues from birth control pills impair nervous system development and swimming behavior of zebrafish larvae. Science of The Total Environment. 2021; 770, 145272.

[2] ANPE 1990. Preliminary diagnosis for the Study of the Ecological Balance of LakeDiagnostic préliminaire pour l'Etude de l'Equilibre Ecologique du lac de Bizerte. GIC-NNEA-TECI, ANPE, TunisiaTunisie

[3] Stegeman JJ. Cytochrome P450 gene diversity and function in marine animals: past, present, and future. Marine Environmental Research. 2000; 50, 61-62.

[4] Ameur WB et al. Polybrominated diphenyl ethers and their methoxylated analogs in mullet (*Mugil cephalus*) and sea bass (*Dicentrarchus labrax*) from Bizerte Lagoon, Tunisia. Marine environmental research. 2011; 72(5), 258-264.

[5] Ameur, WB et al. Organochlorine and organobromine compounds in a benthic fish (*Solea solea*) from Bizerte Lagoon (northern Tunisia): Implications for human exposure. Ecotoxicology and environmental safety, 2013; 88, 55-64.

[6] Ameur, WB et al. Legacy and Emerging Brominated Flame Retardants in Bizerte Lagoon Murex (*Hexaplex Trunculus*): Levels and Human Health Risk Assessment. Archives of environmental contamination and toxicology, 2020; 1-13.

[7] El Megdiche Y et al. Anthropogenic (PBDE) and naturally-produced (MeO-PBDE) brominated compound levels in Bizerte Lagoon clams (*Ruditapes decussatus*): Levels and human health risk assessment. Marine pollution bulletin, 2017; 125(1-2), 176-185.

[8] Mekni S et al. Occurrence of halogenated flame retardants in sediments and sea urchins (*Paracentrotus lividus*) from a North African Mediterranean coastal lagoon (Bizerte, Tunisia). Science of The Total Environment, 2019; 654, 1316-1325.

[9] Mekni S et al. Occurrence of halogenated and organophosphate flame retardants in sediments and eels (*Anguilla anguilla*) from Bizerte Lagoon, Tunisia. Frontiers in Environmental Science, 2020; 8, 67.

[10] Derouiche A et al. Polychlorinated biphenyls in sediments from Bizerte lagoon, Tunisia. Bulletin of environmental contamination and toxicology, 2004; 73(5), 810-817.

[11] Ameur WB et al. Concentration of polychlorinated biphenyls and organochlorine pesticides in mullet (*Mugil cephalus*) and sea bass (*Dicentrarchus labrax*) from Bizerte Lagoon (Northern Tunisia). Chemosphere, 2013; 90(9), 2372-2380.

[12] Barhoumi B et al. Distribution and ecological risk of polychlorinated biphenyls (PCBs) and organochlorine pesticides (OCPs) in surface sediments from the Bizerte lagoon, Tunisia. Environmental Science and Pollution Research, 2014; 21(10), 6290-6302.

[13] Trabelsi S, Driss MR. Polycyclic aromatic hydrocarbons in superficial coastal sediments from Bizerte Lagoon, Tunisia. Marine Pollution Bulletin, 2005; 50(3), 344-348.

[14] Barhoumi B et al. Polycyclic aromatic hydrocarbons (PAHs) in surface sediments from the Bizerte Lagoon, Tunisia: levels, sources, and toxicological significance. Environmental monitoring and assessment, 2014; 186(5), 2653-2669.

[15] Salem FB et al. Distribution of organic contamination of sediments from Ichkeul Lake and Bizerte Lagoon, Tunisia. Marine pollution bulletin, 2017; 123(1-2), 329-338.

[16] Viarengo A et al. The use of biomarkers in biomonitoring: a 2-tier approach assessing the level of pollutant induced stress syndrome in sentinel organisms. Comparative Biochemistry and Physiology Part C Toxicology & Pharmacology. 2007; 146, 281-300.

[17] De Lafontaine Y et al. Biomarkers in zebra mussels (*Dreissena polymorpha*) for the assessment and monitoring of water quality of the St Lawrence river (Canada). Aquatic Toxicololgy. 2000; 50, 51-71.

[18] Mosleh YY et al. Effects of chitosan on oxidative stress and metallothioneins in aquatic worm *Tubifex tubifex* (Oligochaeta, Tubificidae). Chemosphere, 2007; 67, 167-175.

[19] Oliveira MM et al. Brain acetylcholinesterase as a marine pesticide biomarker using Brazilian fishes. Marine Environmental Research. 2007; 63, 303-312.

[20] Sunderman Jr. Biochemical indices of lipid peroxidation in occupational and environmental medicine. In: Foa, V., Emmett, E.A., Maroni, M., Colomi, A. (Eds.), Occupational and Environmental Chemical Hazards: Cellular and Biochemical Indices for Monitoring Toxicity. Wiley, Interscience, New York, pp. 1987; 151-158.

[21] Wong M, Poon B. Sources, fates and effects of persistent organic pollutants in China, with emphasis on the Pearl River Delta, in: H. Fiedler (Ed.), The Hand Book of Environmental Chemistry, Persistent Organic Pollutants, 3rd edn, Springer-Verlag, Berlin/Heidelberg, 2003, Chapter 13.

[22] Essid N, Aissa P. Quantitative study of free nematodes in the North and East sectors of the Bizerte lagoon (Tunisia). Etude quantitative des nématodes libres des secteurs Nord et Est de la lagune de Bizerte (Tunisie). 2002.

[23] Sakka Hlaili A et al. Winter-summer variation of the phytoplankton community of theVariation hiverno-estivale de la communauté phytoplanctonique de la lagune de Bizerte lagoon in natural environments and fertilized with nutrients. Review of the Faculty ofen milieux naturel et fertilisé en nutriments. Revue de la Faculté des Sciences ofde Bizerte, Tunisia, 2003; 2, 37-49.

[24] Tolosa I, Mesa-Albernas M, Alonso-Hernandez CM. Organochlorine contamination PCBs, DDTs, HCB, HCHs in sediments from Cienfuegos bay, Cuba. Marine Pollution Bulletin, 2010; 60, 1619-1624.

[25] Montuori P et al. Polychlorinated biphenyls and organochlorine pesticides in Tiber River and estuary: occurrence, distribution and ecological risk. Science of the Total Environment, 2016, 571, 1001-1016.

[26] Aznar-Alemany Ò et al. Occurrence of halogenated flame retardants in commercial seafood species available in European markets. Food and Chemical Toxicology, 2017;104:35-47.

[27] Weijs L et al. Levels and profiles of persistent organic polluants in several tissues of harbor porpoises (*Phoconea phoconea*) from the Black Sea. Organohalogen Compound, 2009; 71:432-436

[28] Darnerud PO. Toxic effects of brominated flame retardants in man and wildlife. Environment International, 2003; 29, 841-853.

[29] Gill U et al. Polybrominated diphenyl ethers: human tissue levels and

toxicology. Reviews of Environmental Contamination and Toxicology, 2004. 183, 55e97.

[30] Cheikh M, Derouiche A, Driss MR. Determination byDétermination par (CPG ECD) of organochlorine pesticide residues in thedes résidus de pesticides organochlorés dans les sédiments de la lagune de Bizerte lagoon sediments.. Bulletin of thede l'Institut National Institute of Science and Technology of the Seades Sciences et Technologies de la Mer, 2002; 7, 160-163.

[31] Covaci A et al. Levels and distribution of organochlorine pesticides, polychlorinated biphenyls and polybrominated diphenyl ethers in sediments and biota from the Danube Delta, Romania. Environmental Pollution, 2006; 140, 136-149.

[32] de Mora S et al. Distribution of petroleum hydrocarbons and organochlorinated contaminants in marine biota and coastal sediments from the ROPME Sea Area during 2005. Marine Pollution Bulletin, 2010; 60, 2323-2349.

[33] Ennaceur S, Gandoura N, Driss MR. Distribution of polychlorinated biphenyls and organochlorine pesticides in human breast milk from various locations in Tunisia: levels of contamination, influencing factors, and infant risk assessment. Environmental Research, 2008; 108, 86-93.

[34] Ennaceur S, Driss MR. Serum organochlorine pesticide and polychlorinated biphenyl levels measured in delivering women from different locations in Tunisia. International Journal of Environmental Analytical Chemistry, 2010; 90, 821-828.

[35] Necibi M et al. Distributions of organochlorine pesticides and polychlorinated biphenyl in surface water from Bizerte Lagoon, Tunisia. Desalination and Water Treatment, 2015; 56(10), 2663-2671.

[36] Young LY, Cerniglia CE. Microbial Transformation and Degradation of Toxic Organic Chemicals. Wiley, New York. 1995.

[37] Gearing PJ et al. Partitioning of No. 2 fuel oil in controlled estuarine ecosystem, sediments and suspend particulate matter. Environmental Science and Technology, 1980; 14, 1129-1135.

[38] Chiou CT, McGroddy SE, Kile DE. Partition characteristics of polycyclic aromatic hydrocarbons on soils and sediments. Environmental Science and Technology, 1998; 32, 264-269.

[39] Karickhoff S. Semiempirical estimation of sorption of hydrophobic pollutants on natural sediments and soils. Chemosphere, 1981; 10, 833-846.

[40] Barhoumi B et al. Occurrence of polycyclic aromatic hydrocarbons (PAHs) in mussel (*Mytilus galloprovincialis*) and eel (*Anguilla anguilla*) from Bizerte lagoon, Tunisia, and associated human health risk assessment. Continental Shelf Research, 2016; 124, 104-116.

[41] Wang W et al. Depth-distribution, possible sources, and toxic risk assessment of organochlorine pesticides (OCPs) in different river sediment cores affected by urbanization and reclamation in a Chinese delta. Environmental Pollution, 2017; 230, 1062-1072.

[42] Xu X et al. Associations of serum concentrations of organochlorine pesticides with breast cancer and prostate cancer in U.S. adults. Environmental Health Perspectives, 2010; 118, 60-66.

[43] Necibi M, Mzoughi N. Distribution of organochlorine pesticides in sediment

cores from the Bizerte Lagoon (Tunisia). International Journal of Environmental Analytical Chemistry, 2020; 100(10), 1118-1132.

[44] Louati H et al. The roles of biological interactions and pollutant contamination in shaping microbial benthic community structure. Chemosphere, 2013; 93(10), 2535-2546.

[45] Louati H et al. Responses of a free-living benthic marine nematode community to bioremediation of a PAH mixture. Environmental Science and Pollution Research, 2015; 22(20), 15307-15318.

[46] Othman HB et al. Structural and functional responses of coastal marine phytoplankton communities to PAH mixtures. Chemosphere, 2018; 209, 908-919.

[47] Sellami B et al. Effects of anthracene on filtration rates, antioxidant defense system, and redox proteomics in the Mediterranean clam *Ruditapes decussatus* (Mollusca: Bivalvia). Environmental Science and Pollution Research, 2015; 22(14), 10956-10968.

[48] Sellami B et al. The effects of anthracene on biochemical responses of Mediterranean mussels *Mytilus galloprovincialis*. Chemistry and Ecology, 2017; 33(4), 309-324.

[49] Dellali M et al. Multi-biomarker approach in *Mytilus galloprovincialis* and *Ruditapes decussatus* as a predictor of pelago-benthic responses after exposure to Benzo [a] Pyrene. Comparative Biochemistry and Physiology Part C: Toxicology & Pharmacology, 2021; 249, 109141.

[50] Nasri A et al. Using meiobenthic taxa, nematofauna biological traits, and bacterial abundance to assess the effects of the polybrominated diphenyl ethers compound: Case study of tetrabromo diphenyl ether BDE-47. Science of The Total Environment, 2021; 770, 145251.

[51] Nasri A et al. Ecotoxicity of polybrominated diphenyl ether (BDE-47) on a meiobenthic community with special emphasis on nematodes: Taxonomic and trophic diversity assessment. Environmental Pollution, 2021; 277, 116727.

[52] Bodin N et al. PCB contamination in fish community from the Gironde estuary (France): blast from the past. Chemosphere, 2014; 98, 66-72.

[53] Scaps P. A review of the biology, ecology and potential use of the common ragworm *Hediste diversicolor* (O.F. Muller) (Annelida: *Polychaeta*). Hydrobiologia, 2002; 470,203-218.

Section 2

Technologies for POPs Treatment

Chapter 5

Type, Sources, Methods and Treatment of Organic Pollutants in Wastewater

Poslet Shumbula, Collet Maswanganyi and Ndivhuwo Shumbula

Abstract

Persistent organic pollutants (POPs), which are synthetic organic chemical compounds, either intentionally or unintentionally produced, have widely aroused public concern in recent years. These chemicals are toxic and major environmental concern due to their persistence, long range transportability, bioaccumulation and potentially adverse effects on living organisms. Uncontrolled inputs combined with poor environmental management often result in elevated levels of persistent organic pollutants in affected estuaries. Since the Stockholm Convention on POPs was adopted, different techniques have been extensively developed. A major focus revealed the need for low cost methods that can be implemented easily in developing countries such as electrochemical techniques. Persistent organic pollutants are known to be resistant to conventional treatment methods such as flocculation, coagulation, filtration and oxidant chemical treatment. However, various advanced wastewater treatment technologies such as, activated carbon adsorption, biodegradation using membrane bioreactor and advanced oxidation processes (AOPs) have been applied in the treatment of POPs.

Keywords: persistent organic pollutants, environmental contaminants, dioxins, biodegradation, wastewater treatment

1. Introduction

In the past decades, the health effects of environmental pollution on the population have been a growing source of worry around the world. According to the WHO (World Health Organization), one-third of the diseases afflicting humanity are caused by extended exposure to pollution. Since World War II, scientists have identified a number of chemical contaminants that are toxic, persistent in the environment, bioaccumulative, and prone to long-range atmospheric transboundary migration and deposition, and are expected to have serious health consequences for humans, wildlife, and marine biota both near and far from their source of emission. These toxins are chemical contaminants, also called the dirty dozen [1]. Being volatile substances, POPs evaporate into the air in warm regions of the globe, are transported by air currents up to cold regions and in mountainous regions where they condense [2, 3].

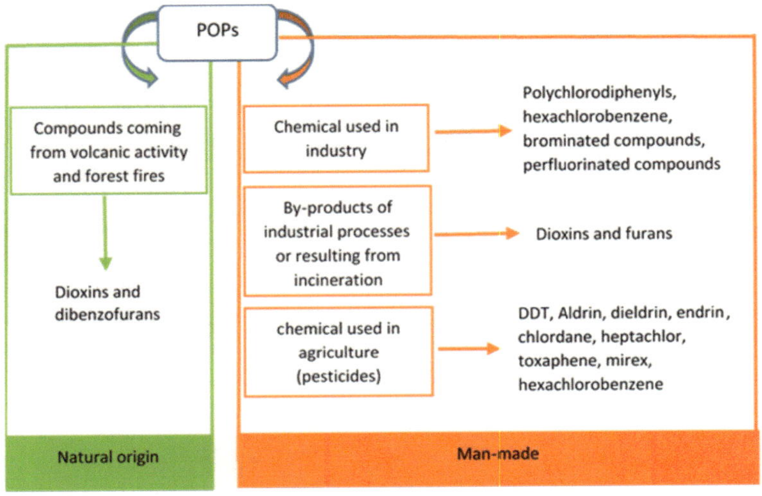

Figure 1.
Classification of persistent organic compounds according to their origin. Picture adapted from [9].

Most POP chemicals are non-polar organic compounds, consequently hydrophobic, with extremely low water solubility. In marine and terrestrial systems, they bind strongly to solids, particularly organic matter, evading the aqueous segment [4]. They are also lipophilic, which means that they accumulate in the fatty tissue of living animals and human beings. The stockpiling in fatty tissue allows the compound to persevere in biota, where the metabolism rate is low [5–8]. Due to the bioaccumulation and biomagnification phenomena, the POP concentration may be much higher in the tissues of the organisms (up to 70,000 higher concentrations). POP concentrations tend to rise as you travel up the food chain, therefore species at the top of the food chain, such as fish, predatory birds, mammals, and humans, have the largest concentrations of these chemicals and are thus at the greatest danger of acute and chronic harmful effects. POPs are mostly man-made chemical products intended to be used in various areas, for an example, in agriculture and industry, or unintentional by-products resulting from industrial processes, or from waste incineration. Different classes of POPs substances such as organochlorinated pesticides (OCP), polychlorinated biphenyl (PCBs), perfluorinated compounds (PFCs), brominated compounds (BFR), dioxins and furans are known. Most of these substances are anthropogenic origin. However, substances such as dioxins and furans may have natural origin (**Figure 1**), such as volcanic activities and vegetation fires [10–17].

2. Types of POPs

Many POPs were widely used during industrial revolution after World War II. However, many of these chemicals proved to be beneficial in pest and disease control, but they had unforeseen effects on human health and environment.
In Stockholm 2001, representatives from 92 countries have agreed to sign the Stockholm Convention on POPs to reduce and/or eliminate the release of 12 original POP substances. More contaminants have been discovered; the main concern is over the original 12. These contaminants are the 10 intentionally produced chemicals: aldrin, endrin, chlordane, DDT, dieldrin, heptachlor, mirex, toxaphene, hexachlorobenzene (HCB) and PCBs and the two unintentionally produced substances polychlorinated dibenzo-p-dioxins (PCDDs) and polychlorinated dibenzofurans

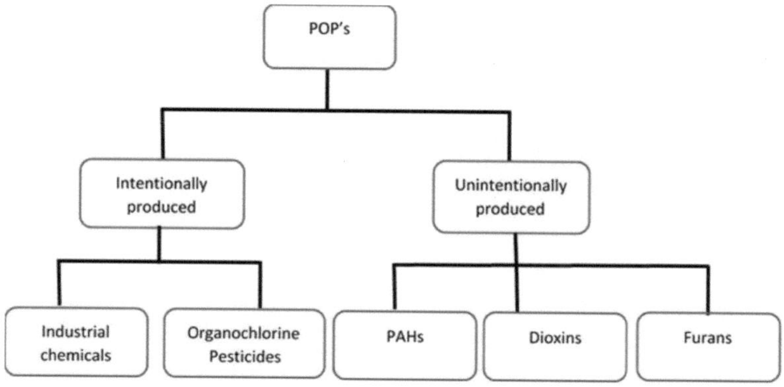

Figure 2.
Categories of persistent organic pollutants. Picture adapted from [20].

(PCDFs) [18, 19]. Another type of interest also classified as persistent organic compounds is polycyclic aromatic hydrocarbons (PAHs). Combustion and burning of organic compounds produces these substances unintentionally. Their occurrences are related to anthropogenic processes, and contamination of PAHs in river sediment is especially serious in high-density industrial areas [18]. Persistent organic pollutants (POPs) are a group of chemicals that have been intentionally or unintentionally produced, and introduced into the environment as shown in **Figure 2**.

2.1 Intentional persistent organic pollutants

Intentionally produced chemicals currently or were once used in agriculture, manufacturing, disease control or industrial processes. These intentional POPs compounds, shown in **Figure 3**, will be produced as wanted products by different chemical reactions that include chlorine. These types are organic molecules with linked chlorine atoms, high lipophilicity and, usually, high neurotoxicity, and they are called organochlorine compounds. Some of the well-known examples of organochlorine compounds are the chlorinated insecticides, such as dichlorodiphenyltrichloroethane, and polychlorinated biphenyls. They have several compounds which can be divided into two types that are industrial chemicals and organochlorine pesticides [22, 23].

2.1.1 Industrial chemicals

Polychlorinated biphenyls, very stable mixtures that are resistant to extreme temperature and pressure, are a group of manmade chemicals, oily liquids or solids, clear to yellow in color, with no smell or taste. They have been discovered in water, sediments, avian tissue, and fish tissue all throughout the planet. These chemicals make up a significant subset of special wastes. PCBs are a group of chemical compounds in which the biphenyl molecule has 2–10 chlorine atoms linked to it. When explaining PCBs, monochlorinated biphenyls (i.e., one chlorine atom bonded to the biphenyl molecule) are frequently mentioned. The chemical structure of chlorinated biphenyls is depicted in **Figure 4**. There are 209 distinct PCB congeners in theory. Many of them are resistant to degradation, allowing them to survive for lengthy periods of time in the environment and spread via air and water transport mechanisms [25–27].

Many industrial applications, such as fire-resistant transformers and insulating condensers, relied heavily on PCBs. Prior to 1977, they were utilized as heat exchanger fluids and in the fabrication of aluminum, copper, iron, and steel [27]. Apart from

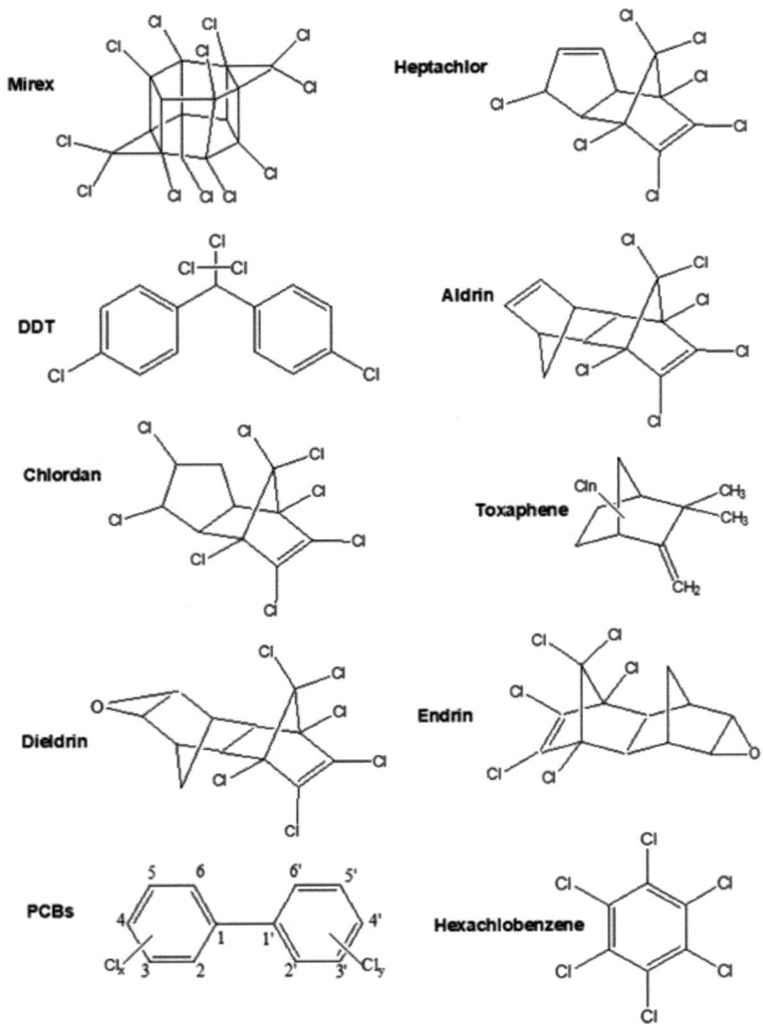

Figure 3.
Intentionally persistent organic pollutants chemical structures. Adapted from [21].

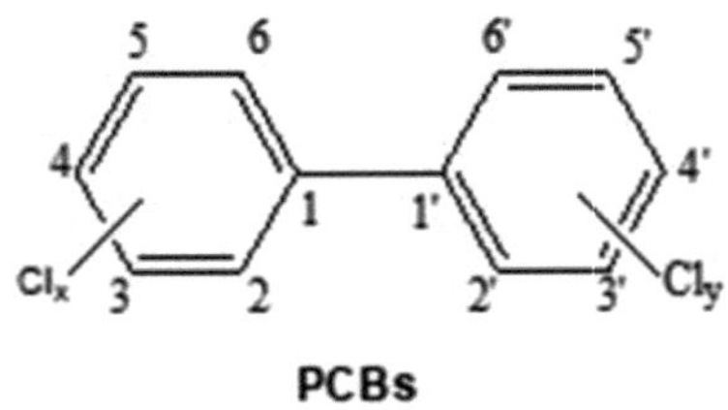

Figure 4.
Industrial POPs chemical structure. Adapted from [24].

their usage in the above applications, they were also applied as plasticizers in natural and synthetic rubber products, as well as adhesives, insulating materials, flame retardants, lubricants in the treatment of wood, clothing, paper, and asbestos, chemical stabilizers in paints and pigments, and as dispersing agents in aluminum oxide formulations. PCBs are frequently discovered in the effluent and sludge of municipal wastewater treatment plants. Although prohibited in the 1980s, PCBs are presently employed in transformers in some parts of the world, especially Brazilian [28, 29].

2.1.2 Organochlorine pesticides

Organochlorine (OC) pesticides are typically man-made synthetic pesticides widely used all over the world. They belong to the group of chlorinated hydrocarbon derivatives, which have vast application in the chemical industry and in agriculture. Pesticides are a class of chemicals used to kill insects, weeds, fungi, bacteria, and other organisms. Insecticides, fungicides, bactericides, herbicides, and rodenticides are some of the terms used to describe them. The majority of pesticides may kill a wide range of pests and weeds, but some are targeted at specific pests or pathogens. Although these substances are typically man-made, plant derivatives and naturally occurring inorganic minerals are examples of exceptions that occur naturally. Since the first naturally occurring pesticide, nicotine derived from tobacco leaf extracts, was employed to control the plum curculio and the lace bug in the seventeenth century. Many chlorinated hydrocarbon insecticides were created in the 1940s, although they were not widely used until the 1950s. Aldrin, dieldrin, heptochlor, and endrin form part of the reported chlorinated hydrocarbon insecticides. However, in spite of their early promise, these organochlorine insecticides are now much less used because of their environmental pollution impact [30, 31].

Pesticides are employed for many different purposes. Pesticide use has increased due to increased agricultural production, resulting in increased pollution of environmental compartments such as soil, water, and air. Pesticide properties like high lipophilicity, bioaccumulation, long half-life, and potential for long-range transport have enhanced the risk of contamination in air, water, and soil, even after many years of use. This occurrence has the potential to become a long-term hazard to the ecosystem's plant and animal groups' coexistence. Pest problems result in the loss of nearly a third of the world's agricultural productivity each year, despite the fact that pesticide consumption exceeds two millio liters each year. A study by Pimentel showed that only a small percentage (0.3%) of applied pesticides goes into the target pest while 99.7% go somewhere else into the environment [32].

Although the application of organochlorine pesticides has been forbidden for a considerable period in many countries, the residues continue to induce a significant impact on the environment and its ecosystems [33]. Overuse or misuse of pesticides has a negative impact on environmental health as well as ecosystem services. Many aquatic and terrestrial animals, have been documented to be toxicated by pesticides. Pesticides have a negative impact on aquatic ecosystems, including microbes, animals, plants, and fish [34–38].

During the last three or four decades, insecticide manufacturing has been rather constant. Insecticides and fungicides, on the other hand, are the most important pesticides for human exposure in food since they are sprayed just before or after harvesting. Herbicide output has risen as chemicals have increasingly supplanted land cultivation in weed management, accounting for the majority of agricultural pesticides. Large amounts of pesticides have the ability to enter water either directly, as in mosquito control applications, or indirectly, as in drainage of agricultural lands [39–41].

DDT was widely employed during World War II to protect soldiers and civilians from malaria, typhoid, and other diseases caused by insects before its insidious effects on humans and wildlife were discovered. DDT was employed to manage disease after the war, and it was sprayed on a number of agricultural crops, particularly cotton. It did the job, reducing the threat of malaria and the loss of income to the agriculture industry [42]. DDT continues to be applied against mosquitoes in several countries to control malaria. Its stability, its persistence, and its widespread use have meant that DDT residues can be found everywhere; residual DDT has even been detected in the Arctic.

2.2 Unintentionally POPs

Unintentionally produced chemicals (see **Figure 5**) are a result of combustion of medical waste, incarnation and some industrial processes. They are divided into three types, viz., polycyclic aromatic hydrocarbons (PAHs), dioxin and furan compounds.

2.2.1 Polycyclic aromatic hydrocarbons (PAHs)

PAHs are ubiquitous group of several hundreds of chemicals that comprise two or more fused benzene rings in linear, angular or cluster arrangements, containing only carbon and hydrogen. The central molecular structure is held together by stable carbon-carbon bonds. They are mostly caused by incomplete combustion of natural or man-made fuels such as coal and wood, as well as vehicular pollutants and cigarette smoke [44]. Dietary exposure accounts for more than 70% of human exposure in non-smokers [45]. According to a dietary survey conducted in the United Kingdom, cereals and oils/fats account for a significant portion of PAH intake [46]. Typical PAH contamination occurs when food is subjected to combustion products in technical procedures such as direct fire drying [47]. High PAH concentrations in charcoal grilled/barbecued foods may also result from certain traditional home cooking methods such as grilling, roasting, frying, and smoking [48]. However, the greatest amounts of PAHs released into the environment are via anthropogenic processes like fossil fuel combustion and by-products of industrial processing. The Environmental Protection Agency (EPA) of the United States included 16 PAHs on a priority pollutants list because they are considered potential or probable human carcinogens. As a result, their dispersal and the likelihood of human exposure have received a lot of interest. PAHs have been found in soil, air, and sediments, as well as on a variety of food and beverage products [49–51].

2.2.2 Dioxins and dibenzofurans

Polychlorinated dibenzo-*p*-dioxins (PCDDs), dibenzofurans (PCDFs) and polychlorinated biphenyls (PCBs) constitute three groups of relevant persistent organic pollutants with enhanced chronic toxicity. PCDD/Fs (**Figure 6**) are emitted by a

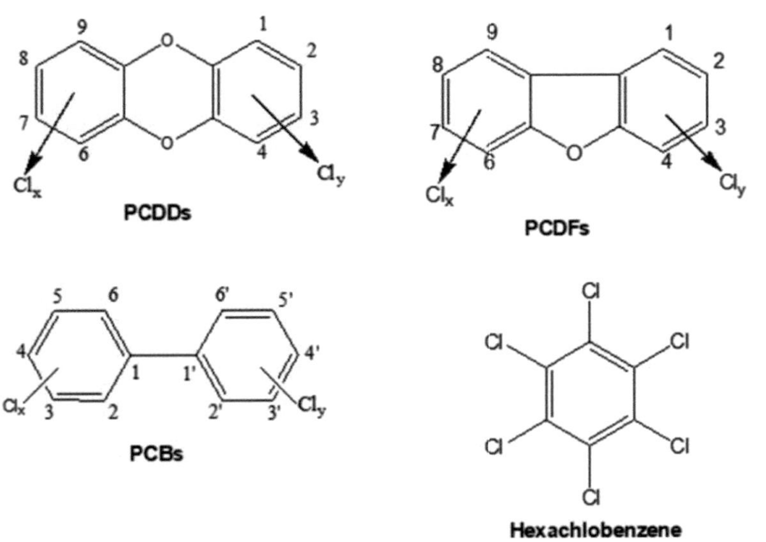

Figure 5.
Unintentional produced POPs chemical structures. Adapted from [43].

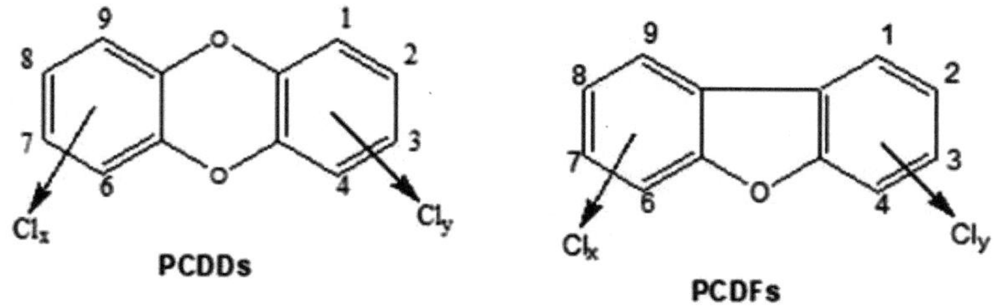

Figure 6.
Molecular structures of chlorinated dibenzo-p-dioxins (dioxins, PCDDs) and dibenzofurans (furans, PCDFs). Adapted from [52].

variety of human activities and industrial processes, and can be referred as undesirable by-products. PCBs are ubiquitous environmental pollutants as a result of their large-scale manufacture till the end of the 1980s and their continued use. PCDD/Fs and PCBs can also be released from stationary sources such as waste incineration and biomass and fossil fuel combustion. PCDD/Fs and PCBs can be considered environmental markers of anthropogenic activities in light of this information, as their occurrence is invariably linked to human activities. PCDDs and PCDFs, commonly called "dioxins", are two classes of "quasi-planar" tricycles aromatic ethers with 210 different compounds (congeners) in total [53].

PCDDs and PCDFs are solids at room temperature and have a rather low volatility. Dispersion in the atmosphere is thus likely to occur mainly in particulate aerosols. The PCDD/F have been of concern for decades because of their toxic properties. A structurally similar series of compounds, the chlorinated dibenzofurans (furans), have similar chemical properties and toxic effects. The most toxic and most extensively studied representative of the chlorinated dioxins (PCDDs) is 2,3,7,8-tetrachlorodibenzo-*p*-dioxin (TCDD). In recent years there has been a growing trend to include a specific subgroup of PCBs, the so-called dioxin-like PCBs (**Figure 5**) which has finally been added to methods along with the dioxins and furans. It is widely acknowledged that man-made sources and activities contribute far more to the environmental burden of PCDDs and PCDFs than natural processes, particularly since the 1930s, when environmental levels have steadily increased in tandem with the large-scale production and use of chlorinated chemicals [54, 55]. Chemical processes, combustion processes, and secondary sources are the three primary categories of man-made sources of PCDDs and PCDFs [56].

3. Sources of POPs

In the past decades, many reports on the dependents of POPs by industry and agricultural sectors were seen. POPs proved to be beneficial in pest and disease control, crop production, and industrial applications. Many were widely used commercially during the boom in industrial production after World War II, resulting in wide geographical distribution. **Figure 7** shows some of the sources related to POPs [57].

POPs are extremely stable in all environmental elements. They are discharged into the atmosphere through a variety of industrial sources, including power plants, heating plants, and incinerating facilities, as well as from domestic furnaces, transportation, agricultural sprays, and evaporation from water surfaces, soil, and landfills. Other sources of POPs compounds, such as inadvertent generation, can be present in incinerations, chemical plants, other combustions, forest fires, putrefaction, and PCB-containing wastes. This type of trash can be found in a variety of

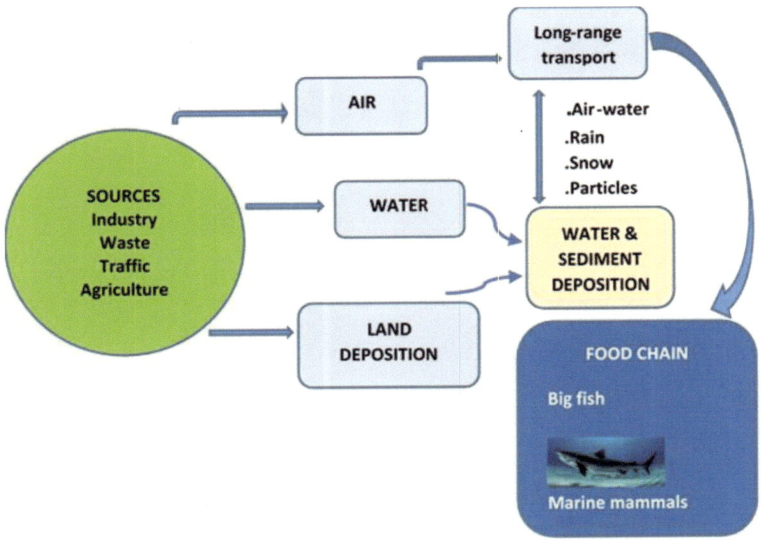

Figure 7.
Schematic depicting POPs in the environment and main environmental processes during long-distance atmosphere transport, bioaccumulation, and biomagnification. Adapted from [57].

places and stems from a variety of activities, such as the use of old oil, equipment repair and maintenance, and building destruction [58, 59].

Wastewaters from plants generating or using POPs, as well as runoff from fields and roads, and atmospheric deposition, are the origins of pollutants, oil, fates, liquid fuels, dirt, ash, and silt entering the water system. Oceans and seas are their greatest reservoirs, where they collect from river sediments, air deposition, trash disposal, and accidents. They are retained in sediments on the bottoms of seas,

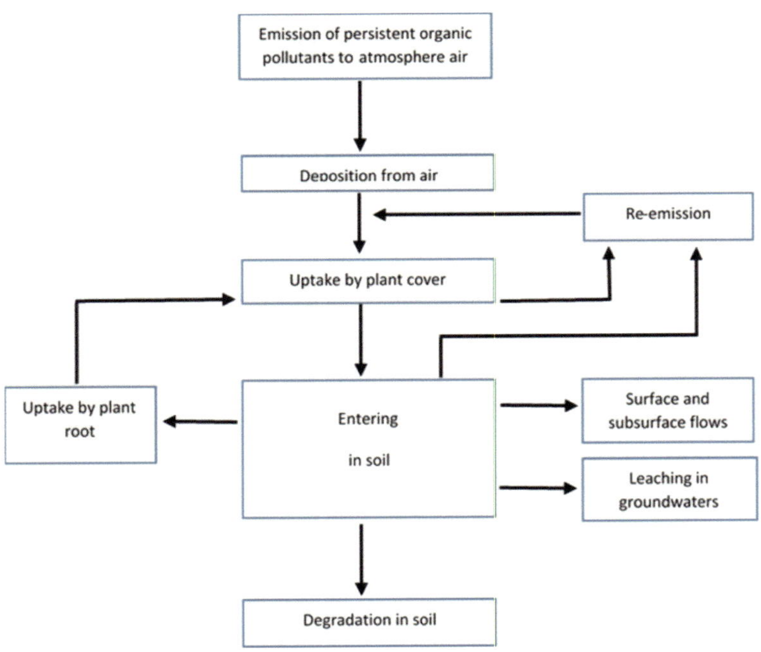

Figure 8.
Conceptual model for the behavior of persistent organic pollutants in the air-plant soil system. Adapted from [60].

oceans, and huge lakes, where they can be released and re-enter the atmosphere after a period of time, as indicated in **Figure 8** [42, 60].

4. Methods for treatment of persistent organic pollutants

POPs have adversely posed a health concern worldwide for ages. Due to their concerning health issues, some countries have resorted to reduce the use of chemicals or processes that produce POPs, while others have prohibited them entirely. However, most processes that result in the production of POPs are beneficial to both human and environmental health. This is because some POPs are produced during production of synthetic chemicals for crop production, medication, clothing etc. In addition, some POPs are unavoidable since they may be produced unintentional from simple combustions. Since most of these POPs end up in water streams, various methods for wastewater treatment have been implement and reported in literature. These methods ensure the conversion of wastewater into portable water by removing harmful and hazardous chemicals [61]. Conventional treatment refers to some of the most effective water treatment procedures used in the service and distribution of industrial or municipal potable water. At different stages of a typical treatment method, any of the physical, chemical, or biological channels provide good combination.

Preliminary, primary, secondary, and tertiary wastewater treatment stages are in sequence of increasing treatment level, with final pH adjustments as needed. The chosen conventional approach must be able to meet the regulatory authority's recommended microbiological and chemical criteria while operating and maintaining at a low cost [62].

Conventional treatment methods such as flocculation, coagulation, filtration, and oxidant chemical treatment are ineffective against POPs. The chemical properties of POPs, such as, low water and high fat solubility, stability to all degradation processes and low vapor pressure, are the main components for their efficiency as pesticides and for their persistence in the environment [63]. The inability in some instances to remove POPs from wastewater using conventional methods have prompted scientists to develop other methods. Various advanced wastewater treatment technologies such as, activated carbon adsorption, biodegradation using membrane bioreactor [64] and advanced oxidation processes [65] have been applied in the treatment of POPs. This is because of growing number of emerging POPs that are being identified in water and the concerns that are accompanied by human and environmental health hazard [66]. Various setbacks such as cost, sophisticated instrumentation, low degradation efficiency, generation of toxic secondary chemicals and massive sludge production have recently been addressed using advanced methods and technologies. Below is the short discussion of biodegradation and advanced oxidation processes wastewater treatment technologies.

4.1 Biodegradation

Biodegradation is an evolving technology that comprises the application of selected living microorganisms to degrade, metabolize/immobilize any unwanted substances such as pesticides, organic pollutants and hydrocarbons from soil and water, to improve its quality [67]. Although every microorganism has the ability to eradicate pollutants, only few particular or engineered microorganisms are used broadly to eradicate pollutants efficiently. Bioremediation technology, applied in perspective to POPs removal, takes into consideration the following methods: (1) bioventing: aerating water to stimulate *in situ* biodegradation of organic contaminants and promote bioremediation, (2) biostimulation: modification of contaminated media to provide the nutrition to soil microbiota by adjusting pH, addition of limiting nutrient to improve C: N: P ratio, and (3) bioaugmentation: addition of microbial community (bacteria and fungi) and any biocatalyst (gene and enzyme) to degrade organic/inorganic pollutants [68]. One of the most important variables in the efficient breakdown of petrochemical wastes in a given ecosystem is microorganism selection. It is because only those microbial species are adapted to work in that specific habitat. Likewise, intermediates created during photocatalytic degradation processes are harmful to a variety of creatures in the environment [69].

Currently, the membrane bioreactor does not always achieve the desired results in the treatment of POPs, and it performs poorly in the removal of non-biodegradable aliphatic and aromatic hydrocarbon compounds, halogenated organic compounds, organic dyes, pesticides, and phenols and their derivatives. The process technicalities and economic feasibilities are the two most significant assessment elements for achieving the goal in wastewater treatment technology [70].

4.2 Advanced oxidation processes

The use of conventional methods is not wholly accepted nowadays because of the high costs and operational problems. Consequently, it is necessary to adopt modern systems like advanced oxidative processes (AOPs) [71]. Some of the AOPs' characteristics include: (1) potential capacity for mineralization of organic pollutants to carbon dioxide and water, as well as oxidation of inorganic compounds and ions such as chlorides, nitrates, and others; (2) non-selective reactivity with the vast majority of organic compounds, which is particularly appealing to avoid the presence of potentially toxic by-products from the primary pollutants that can be produced by other methods that do not achieve complete oxidation [65]. Some of the AOPs discussed below.

4.2.1 Catalysts in advanced oxidation processes

AOPs have successfully used both homogeneous and heterogeneous catalysts. Heterogeneous systems have obvious advantages over homogeneous systems, such as the ability to separate the catalyst easily for reuse from the treated water, the lack of a secondary treatment to remove dissolved metals from the treated water, and the ability to withstand extreme operating conditions. The system is also effective over a broader pH range including the common pH for natural water and wastewater (pH 2–9) [72].

The AOPs, as water treatment processes, are performed at pressure and temperature close to environmental conditions. They involve the generation of hydroxyl radicals in sufficient quantity to interact with the organic compounds of the medium. Hydroxyl radicals are the best of the powerful oxidants because they meet a number of criteria, including: (1) they do not generate additional waste; (2) they are not toxic and have a short lifetime; (3) they are not corrosive to equipment; and (4) usually produced by easy-to-manipulate assemblies [73]. The following are some of the most common approaches used for this purpose: UV alone, UV/H_2O_2, UV/Fe^{3+}, UV/H_2O_2/Fe^{3+}, UV/O_3, UV/$S_2O_8^{2-}$, UV/TiO_2, UV/chlorine and UV in combination with other photocatalysts. The major issue is the removal efficiency of specific target contaminants by the UV AOPs. UV AOP removal rates vary depending on the molecular structure of the pollutants, both in terms of direct photolysis and radical processes. Furthermore, water matrix effects have a significant influence on removal rates. As a result, each UV AOP system must be individually controlled in line with its water matrix and targeted contaminant removal for optimal POPs control [74]. In most situations, the UV/chlorine oxidation process outperformed UV alone or chlorination, according to Xiang et al. [75]. During the UV/chlorine reaction, hydroxyl and Cl radicals were produced, with the hydroxyl radical taking the lead in the oxidation process. Its contribution to the rate of diuron degradation was calculated to be 28.95%.

4.2.2 Photo-Fenton oxidation

Most AOPs use a combination of oxidants and irradiation (O_3/H_2O_2/UV) or a catalyst and irradiation (Fe^{2+}/H_2O_2; UV/TiO_2) to achieve their goals. The disadvantages that make them economically undesirable vary depending on the AOP are: (1) high electricity demand (for example, ozone and UV-based AOPs), (2) relatively large volumes of oxidants and/or catalysts (for example, ozone, hydrogen peroxide, and iron-based AOPs), and (3) pH operating conditions (e.g. Fenton and photo-Fenton) [76]. Photo-Fenton oxidation system has been identified as a feasible oxidation system for treating these wastewaters. In Fenton and Fenton-like reactions, hydroxyl radicals are usually generated from H_2O_2 catalyzed by iron (Fe^{2+}, α-Fe_2O_3, Fe_3O_4, $H_2Fe_2O_4$, α-FeOOH, etc.) [77]. Nonetheless, the cost effectiveness is one of the major concerns. However, the cost reduction can be obtained through application of heterogeneous catalysts, chelating agent, solar energy and integration with biological treatment technologies [78].

4.2.3 Electrochemical oxidation processes

Electrochemical oxidation procedures, among the numerous AOPs, are gaining popularity for water and wastewater decontamination due to their low cost and high efficiency. Dissolved organic contaminants are primarily oxidized in electrochemical oxidation processes by (i) direct anodic oxidation on the anode surface via charge transfer, and (ii) interaction with physio- and/or chemisorbed hydroxyl radical produced during water oxidation [79]. Electrochemical AOPs have been widely explored for the total degradation of POPs. The electrochemical oxidation is an effective and environmentally friendly technology because it does not require chemicals, only electric current is consumed. The first one is direct oxidation which occurs when the compound reacts directly at the anode's surface or by physisorbed or chemisorbed •OH. The second mechanism is indirect oxidation, which is achieved through the electrochemical generation of a mediator in the bulk solution such as ozone (O_3), hydrogen peroxide (H_2O_2), active chlorine, active bromine or $S_2O_8^{2-}$, among others [80].

Recently, coupling approaches including an electrochemical pre-treatment followed by a biological process have been proposed as cost-effective and reliable remediation methods for persistent chemicals mineralization. This opens the door to more selective electrochemical methods than those involving hydroxyl radicals do, because the goal of the pre-treatment is no longer to achieve total mineralization of non-biodegradable species, but rather to improve their biodegradability by focusing on functional groups that have been shown to reduce biodegradability [81].

4.2.4 Nanofibers

In the one-time elimination of POPs, nanofibers have demonstrated to be the most effective. These adsorbents, on the other hand, demonstrate adaptability in the collection of pollutants. The use of fiber layers with varied pore channels and surface chemistry to produce selectivity for a target chemical could be researched further. Because adsorption is a common water treatment method, the production and operational costs of adsorbent materials are crucial to the introduction of any new classes of materials [82]. Physically and chemically stable carbon-based materials alone (without metals) have also been successfully used as the electro-catalysts [83]. Inexpensive, non-noble transition metals or their oxides supported in carbon nanotube has been reported for treatment of POPs. Bismuth-based nanocomposites [84], copper-reduced graphene oxide electrode [85], boron-doped diamond [86], with different boron and substrate silicon or niobium content [87] have indicated to be an efficient technology for treating POPs wastewater.

5. Conclusions

The POPs are organic compounds of anthropogenic origin, and are resistant to environmental degradation through chemical, biological, or photolytic processes and as a result, accumulate in the food chain. Contamination by POPs is widespread, and circulate globally via the atmosphere, oceans, and other pathways. The Stockholm Convention defines criteria for new POP candidates in terms of their persistence, long-range transport, bioaccumulation and toxicity. Recognizing the dangers of POPs, countries began limiting their production, use, and release. This global, legally binding agreement is to reduce and eliminate the release of 12 POPs, including pesticides and industrial chemicals, as well as unintentionally produced POPs. Conventional water treatment facilities have failed to effectively degrade persistent contaminants from wastewater. However, advanced water treatment options such as activated carbons, membrane bioreactors and advanced oxidation processes are well documented for their capital intensive treatment of these recalcitrant pollutants.

Acknowledgements

The authors would like to thank the University of Limpopo for the financial assistance towards this project.

Conflict of interest

Authors report no conflict of interest.

Author details

Poslet Shumbula[1]*, Collet Maswanganyi[1] and Ndivhuwo Shumbula[2]

1 University of Limpopo, Polokwane, South Africa

2 University of Witwatersrand, Johannesburg, South Africa

*Address all correspondence to: poslet.shumbula@ul.ac.za

References

[1] Ashraf MA, Sarfraz M, Naureen R. Handbook of Environmental Impacts of Metallic Elements: Speciation, Bioavailability and Remediation. Singapore: Springer; 2015. DOI: 10.1007/978-981-287-293-7

[2] Daly GL, Wania F. Organic contaminants in mountains. Environmental Science and Technology. 2005;**39**(2):385-398. DOI: 10.1021/ES048859U

[3] Yang R, Wang Y, Li A, Zhang Q, Jing C, Wang T, et al. Organochlorine pesticides and PCBs in fish from lakes of the Tibetan Plateau and the implications. Environmental Pollution. 2010;**158**(6):2310-2316. DOI:org/10.1016/j.envpol.2010.02.004

[4] Vallack HW, Bakker DJ, Brandt I, Broström-Lundén E, Brouwer A, Bull KR, et al. Controlling persistent organic pollutants-what next? Environmental Toxicology and Pharmacology. 1998;**6**:143-175. DOI: DOI.org/10.1016/S1382-6689(98)00036-2

[5] La Merrill M, Emond C, Kim MJ, Antignac JP, Le Bizec B, Clement K, et al. Toxicological function of adipose tissue: Focus on persistent organic pollutants. Environmental Health Perspectives. 2013;**121**:162-169. DOI: DOI.org/10.1289/ehp.1205485

[6] Yu M, Luo X-J, Wu J-P, Chen S-J, Mai B-X. Bioaccumulation and trophic transfer of polybrominated diphenyl ethers (PBDEs) in biota from the Pearl River Estuary, South China. Environment International. 2009;**35**:1090-1095. DOI: 10.1016/j.envint.2009.06.007

[7] Tomy GT, Palace VP, Halldorson T, Braekevelt E, Danell R, Wautier K, et al. Bioaccumulation, biotransformation, and biochemical effects of brominated diphenyl ethers in Juvenile Lake Trout (*Salvelinus namaycush*). Environmental Science and Technology. 2004;**38**:1496-1504. DOI: DOI.org/10.1021/es035070v

[8] Deribe E, Rosseland BO, Borgstrom R, Salbu B, Gebremariam Z, Dadebo E, et al. Bioaccumulation of persistent organic pollutants (POPs) in fish species from Lake Koka, Ethiopia: The influence of lipid content and trophic position. Science of the Total Environment. 2011;**410**:136-145. DOI: 10.1016/j.scitotenv.2011.09.008

[9] Curtean-Bănăduc A, Lyche JL, Berg V, Burcea A, Bănăduc D. Assessment and Monitoring of Persistent Organic Pollutants in Lotic Ecosystems Methodological Guide. Romania: "Lucian Blaga" University of Sibiu; 2016. ISBN: 978-606-12-1414-3

[10] Ferreira AP. Polychlorinated Biphenyl (PCB) congener concentrations in aquatic birds. Case study: Ilha Grande Bay, Rio de Janeiro, Brazil. Anais da Academia Brasileira de Ciências. 2013;**85**:1379-1388. DOI: 10.1590/0001-3765201398112

[11] Liu W, Chen J, Hu J, Ling X, Tao S. Multi-residues of organic pollutants in surface sediments from littoral areas of the Yellow Sea, China. Marine Pollution Bulletin. 2008;**56**(6):1091-1103. DOI: 10.1016/j.marpolbul.2008.03.021

[12] Hao Q, Sun YX, Xu XR, Yao ZW, Wang YS, Zhang ZW, et al. Occurrence of persistent organic pollutants in marine fish from the Natuna Island, South China Sea. Marine Pollution Bulletin. 2014;**85**(1):274-279. DOI: 10.1016/j.marpolbul.2014.05.058

[13] Zhang Y, Luo XJ, Wu JP, Liu J, Wang J, Chen SJ, et al. Contaminant pattern and bioaccumulation of legacy and emerging organhalogen pollutants in the aquatic biota from an e-waste recycling region in South China. Environmental Toxicology and

Chemistry. 2010;**29**(4):852-859. DOI: 10.1002/etc.122

[14] Tang B, Zeng YH, Luo XJ, Zheng XB, Mai BX. Bioaccumulative characteristics of tetrabromobisphenol A and hexabromocyclododecanes in multi-tissues of prey and predator fish from an e-waste site, South China. Environmental Science and Pollution Research. 2015;**22**:12011-12017. DOI: 10.1007/s11356-015-4463-1

[15] Lana NB, Berton P, Covaci A, Ciocco NF, Barrera-Oro E, Atencio A, et al. Fingerprint of persistent organic pollutants in tissues of Antarctic notothenioid fish. Science of the Total Environment. 2014;**499**:89-98. DOI: 10.1016/j.scitotenv.2014.08.033

[16] Webster, Russell M, Walsham P, Hussy I, Lacaze JP, Phillips L, et al. Halogenated persistent organic pollutants in relation to trophic level in deep sea fish. Marine Pollution Bulletin. 2014;**88**(1):14-27. DOI: 10.1016/j.marpolbul.2014.09.034

[17] Schmid P, Kohler M, Gujer E, Zennegg M, Lanfranchi M. Persistent organic pollutants, brominated flame retardants and synthetic musks in fish from remote alpine lakes in Switzerland. Chemosphere. 2007;**67**(9):S16-S21. DOI: 10.1016/j.chemosphere.2006.05.080

[18] Kendrovski V, Stikova E, Kolevska L. Contamination of Food and Agro Products in the Republic of Macedonia. Archives of Industrial Hygiene and Toxicology. 2001;**52**:69-73

[19] Fiedler H. The Hand Book of Environmental Chemistry, Persistent Organic Pollutants. 3rd ed. Berlin/ Heidelberg: Springer-Verlag; 2003 Chapter 1

[20] Ashraf MA. Persistent Organic Pollutants (POPs): A global issue, a global challenge. Environmental Science and Pollution Research. 2017;**24**(4):4223-4227. DOI: 10.1007/s11356-015-5225-9

[21] Tang HP. Recent development in analysis of persistent organic pollutants under the Stockholm Convention. TrAC Trends in Analytical Chemistry. 2013;**45**:48-66. DOI: 10.1016/j.trac.2013.01.005

[22] Harner T, Pozo K, Gouin T, Macdonald A, Hung H, Cieny I, et al. Global pilot study for persistent organic pollutants (POPs) using PUF disk passive air samplers. Environmental Pollution. 2006;**144**:445. DOI: 10.1016/j.envpol.2005.12.053

[23] Kelce C, Stone S, Laws C, Gray E, Kemppalnen J, Wilson EM. Persistent DDT metabolite p,p'-DDE is a potent androgen receptor antagonist. Nature. 1995;**375**:581. DOI: 10.1038/375581A0

[24] Miniero R, Iamiceli AL. Persistent Organic Pollutants. Italian National Institute of Health: Rome, Italy; 2008. pp. 2672-2682

[25] Villa S, Finizio A, Diazdiaz R, Vighi N. Distribution of organochlorine pesticides in pine needles of an Oceanic Island: The case of Tenerife (Canary Islands, Spain). Water, Air, and Soil Pollution. 2003;**146**:335. DOI: 10.1023/A:1023906306701

[26] Panic O, Gorecki T. Comprehensive two-dimensional gas chromatography (GC x GC) in environmental analysis and monitoring. Analytical and Bioanalytical Chemistry. 2006;**384**:1366. DOI: 10.1007/s00216-006-0568-1

[27] Korytar D, Leonards P, Bore J, Brikmoan U. High-resolution separation of polychlorinated biphenyls by comprehensive two-dimensional gas chromatography. Journal of Chromatography. A. 2002;**958**:203-218. DOI: 10.1016/s0021-9673(02)00327-8

[28] Amato D, Torres J, Malm O. DDT (dichlorodiphenyltrichloroethane): Toxicity and Environmental Contamination: A review. Quimera

Nova. 2002;**25**:995-1002. DOI: 10.1590/s0100-40422002000600017

[29] Amador A. Buried pesticides waste hazard to Poland. Waste Management & Research: The Journal for a Sustainable Circular Economy. 1992;**10**(5):387-398. DOI: 10.1177/0734242X9201000502

[30] Strompl C, Thiele JH. Comparative fate of 1,1-diphenylethylene (DPE), 1,1-dichloro-2,2-bis(4-chlorophenyl)-ethylene (DDE), and pentachlorophenol (PCP) under alternating aerobic and anaerobic conditions. Archives of Environmental Contamination and Toxicology. 1997;**33**:350

[31] Aspelin A. Pesticide Industry Sales and Usage: 1992 and 1993 Market Estimates. U.S: Environmental Protection Agency; 1994

[32] Pimentel D. Amounts of pesticides reaching target pests: Environmental impacts and ethics. Journal of Agricultural and Environmental Ethics. 1995;**8**(1):17-29. DOI: 10.1007/BF02286399

[33] Chen S, Shi L, Shan Z, Hu Q. Determination of organochlorine pesticide residues in rice and human and fish fat by simplified two-dimensional gas chromatography. Food Chemistry. 2007;**104**:1315. DOI: 10.1016/j.foodchem.2006.10.032

[34] Mancini F. Acute pesticide poisoning among female and male cotton growers in India. International Journal of Occupational and Environmental Health. 2005;**11**:221-232. DOI: 10.1179/107735205800246064

[35] Grande M, Andersen S, Berge D. Effects of pesticides on fish. Norwegian Journal of Agricultural Sciences. Experimental and Field Studies. 1994;**13**:195-209

[36] De Lorenzo ME, Scott GI, Ross PE. Toxicity of pesticides to aquatic microorganisms: A review. Environmental Toxicology and Chemistry. 2001;**20**: 84-98. DOI: 10.1002/etc.5620200108

[37] Castillo LE, Martinez E, Ruepert C, Savage C, Gilek M, Pinnock M, et al. Water quality and macroinvertebrate community response following pesticide applications in a banana plantation, Limon, Costa Rica. Science of the Total Environment. 2006;**367**:418-432. DOI: 10.1016/j.scitotenv.2006.02.052

[38] Frankart C, Eullaffroy P, Vernet G. Comparative effects of four herbicides on non-photochemical fluorescence quenching in Lemna minor. Environmental and Experimental Botany. 2003;**49**:159-168. DOI: 10.1016/S0098-8472(02)00067-9

[39] Weinbe J. An overview of POPs treaty-public forum on persistent organic pollutants. The International POPs, Elimination Network. Vol. 8. No. 2. San Francisco: Global Pesticide Campaigner; 1998:9

[40] Junior J, Poppi N. Determination of organochlorine pesticides in ground water samples using solid-phase microextraction by gas chromatography-electron capture detection. Talanta. 2007;**72**:1833-1841. DOI: 10.1016/j.talanta.2007.02.024

[41] Espinosa M, Granada A, Carreno J, Salvatierra M, Serrano F, Olea N. Organochlorine pesticides in placentas from Southern Spain and some related factors. Placenta. 2007;**28**:631. DOI: 10.1016/j.placenta.2006.09.009

[42] El-Shahawi MS, Hamza A, Bashammakh AS, Al-Saggaf WT. An overview on the accumulation, distribution, transformations, toxicity and analytical methods for the monitoring of persistent organic pollutants. Talanta. 2010;**80**:1587-1597. DOI: 10.1016/j.talanta.2009.09.055

[43] Guo W, Pan B, Sakkiah S, Yavas G, Ge W, Zou W, et al. Persistent pollutants

in food: Contamination sources, health effects and detection methods. International Journal of Environmental Research and Public Health. 2019;**16**:4361. DOI: 10.3390/ijerph1624361

[44] Abdel-Shafy HI, Mansour MSM. A review on polycyclic aromatic hydrocarbons: Source, environmental impact, effect on human health and remediation. Egyptian Journal of Petroleum. 2016;**25**:107-123. DOI: 10.1016/j.ejpe.2015.03.011

[45] Martorell I, Perelló G, Martí-Cid R, Castell V, Llobet JM, Domingo JL. Polycyclic aromatic hydrocarbons (PAH) in foods and estimated PAH intake by the population of Catalonia, Spain: Temporal trend. Environment International. 2010;**36**:424-432. DOI: 10.1016/j.envint.2010.03.003

[46] Dennis MJ, Massey RC, McWeeny DJ, Knowles ME, Watson D. Analysis of polycyclic aromatic hydrocarbons in UK total diets. Food and Chemical Toxicology. 1983;**21**:569-574. DOI: 10.1016/0278-6915(83)90142-4

[47] Dennis MJ, Massey RC, Cripps G, Venn I, Howarth N, Lee G. Factors affecting the polycyclic aromatic hydrocarbon content of cereals, fats and other food products. Food Additives and Contaminants. 1991;**8**:517-530. DOI: 10.1080/02652039109374004

[48] Domingo JL, Nadal M. Human dietary exposure to polycyclic aromatic hydrocarbons: A review of the scientific literature. Food and Chemical Toxicology. 2015;**86**:144-153. DOI: 10.1016/j.fct.2015.10.002

[49] Aboul-Kassim T, Simueit B. Aliphatic and aromatic Hydrocarbons in Particulate Fallout of Alexandria, Egypt: Sources and Implications. Environmental Science & Technology. 1995;**29**:2473-2483

[50] Agency for Toxic Substance and Disease Registry (ATSDR). Toxicology Profile for Polycyclic Aromatic Hydrocarbons. Atlanta, GA: US Department of Health and Human Services, Public Health Service; 1995

[51] Garg V. Dioxins and furans: A brief overview. Acta Ciencia Indica. 2017;**43**(1):75-81

[52] Wittich R-M. Degradation of dioxin-like compounds by microorganisms. Applied Microbiology and Biotechnology. 1998;**49**:489-499. DOI: 10.1007/s002530051203

[53] Ferreira AP. Polychlorinated dibenzo-p-dioxins (PCDDs), polychlorinated dibenzofurans (PCDFs), and polychlorinated biphenyls (PCBs) in olivaceous cormorant (*Phalacrocorax brasilianus*) from Sepetiba Bay, Rio de Janeiro, Brazil. 2012;**2**(4):27-33. DOI: 10.5923/j.ms.20120204.01

[54] Jolly J, Jacob C. Review of environmental and human exposure to persistent organic pollutants. Asian Social Science. 2013;**9**(11):107-120. DOI: 10.5539/ass.v9n11p107

[55] Fortin C, Caldbick D. Are dioxins and furans predominantly anthropogenic? Organohalogen Compounds. 1997;**32**:417-429

[56] Alcock R, Gemmill R, Joues K. Improvements to the UK PCDD/F and PCB atmospheric emission inventory following an emissions measurement programme. Chemosphere. 1998;**38**:759. DOI: 10.1016/S0045-6535(98)00222-7

[57] Chandra R, Kumar V. Persistent Organic Pollutants (POPs): Health Hazards and Challenges for their Bioremediation in Environment. 2016. Chapter 1

[58] Ying Y, Yong Y, Dawson R, Yajuan S, Zhang H, Wong T, et al. A spatial temporal assessment of pollution from PCBs in China. Chemosphere.

2005;**60**:731. DOI: 10.1016/j.chemosphere.2005.05.001

[59] Wenzl T, Simon R, Kleiner J, Auklam E. Analytical methods for polycyclic aromatic hydrocarbons (PAHs) in food and the environment needed for new food legislation in the European Union. Trends in Analytical Chemistry. 2006;**25**:716. DOI: 10.1016/j.trac.2006.05.010

[60] Galiulin R, Bashkin V, Galiulina R. Review: Behavior of persistent organic pollutants in the air-plant-soil system. Water, Air, and Soil Pollution. 2002;**137**:179-191. DOI: 10.1023/A:1015558526154

[61] Sharkey M, Harrad S, Abou-Elwafa Abdallah M, Drage DS, Berresheim H. Phasing-out of legacy brominated flame retardants: The UNEP Stockholm Convention and other legislative action worldwide. Environment International. 2020;**144**:106041. DOI: 10.1016/j.envint.2020.106041

[62] Pariatamby A, Kee YL. Persistent organic pollutants management and remediation. Procedia Environmental Sciences. 2016;**31**:842-848. DOI: 10.1016/j.proenv.2016.02.093

[63] Katsoyiannis A, Samara C. Persistent organic pollutants (POPs) in the sewage treatment plant of Thessaloniki, northern Greece: Occurrence and removal. Water research. 2004;**38**:2685-2698. DOI: 10.1016/j.watres.2004.03.027

[64] Baêta BEL, Luna HJ, Sanson AL, Silva SQ, Aquino SF. Degradation of a model azo dye in submerged anaerobic membrane bioreactor (SAMBR) operated with powdered activated carbon (PAC). Journal of Environmental Management. 2013;**128**:462-470. DOI: 10.1016/j.jenvman.2013.05.038

[65] Cuerda-correa EM, Alexandre-franco MF, Fern C. Advanced oxidation processes for the removal of antibiotics from water. An overview. Antibiotics from Water Review. 2020;**12**(1):102. DOI: 10.3390/w12010102

[66] Alharbi O, Basheer AA, Khattab RA, Ali I. Health and environmental effects of persistent organic pollutants. Journal of Molecular Liquids. 2018;**263**:442-453. DOI: 10.1016/j.molliq.2018.05.029

[67] Ojuederie OB, Babalola OO. Microbial and plant-assisted bioremediation of heavy metal polluted environments: A review. International Journal of Environmental Research and Public Health. 2017;**14**:1504. DOI: 10.3390/ijerph14121504

[68] Gaur N, Narasimhulu K, Pydisetty Y. Recent advances in the bio-remediation of persistent organic pollutants and its effect on environment. Journal of Cleaner Production. 2018;**198**:1602-1631. DOI: 10.1016/j.jclepro.2018.07.076

[69] Singh P, Borthakur A. A review on biodegradation and photocatalytic degradation of organic pollutants: A bibliometric and comparative analysis. Journal of Cleaner Production. 2018;**196**:1669-1680. DOI: 10.1016/j.jclepro.2018.05.289

[70] Badmus KO, Tijani JO, Massima E, Petrik L. Treatment of persistent organic pollutants in wastewater using hydrodynamic cavitation in synergy with advanced oxidation process. Environmental Science and Pollution Research International. 2018;**25**(8):7299-7314. DOI: 10.1007/s11356-017-1171-z

[71] Garrido-cardenas JA, Agüera A. Wastewater treatment by advanced oxidation process and their worldwide research trends. International Journal of Environmental Research and Public Health. 2020;**17**:170. DOI: 10.3390/ijerph17010170

[72] Kurian M. Advanced oxidation processes and nanomaterials—A review.

Cleaner Engineering and Technology. 2021;**2**:100090. DOI: 10.1016/j.clet.2021.100090

[73] Garcia-Segura S, El-Ghenymy A, Centellas F, Rodríguez R, Arias C, Garrido J, et al. Comparative degradation of the diazo dye direct yellow 4 by electro-fenton, photoelectro-fenton and photo-assisted electro-fenton. Journal of Electroanalytical Chemistry. 2012;**681**:36-43. DOI: 10.1016/j.jelechem.2012.06.002

[74] Wols BA, Hofman-caris CHM. Review of photochemical reaction constants of organic micropollutants required for UV advanced oxidation processes in water. Water Research. 2012;**46**(9):2815-2827. DOI: 10.1016/j.watres.2012.03.036

[75] Xiang H, Gao N, Lu X, An N, Tan C, Zheng Z. Degradation of diuron by chlorination and UV/chlorine process: Degradation kinetics and the formation of disinfection by-products. Separation and Purification Technology. 2018;**202**:365-372. DOI: 10.1016/j.seppur.2018.03.073

[76] Klamerth N, Rizzo L, Malato S, Maldonado MI, Agüera A, Fernández-Alba AR. Degradation of fifteen emerging contaminants at microg L(−1) initial concentrations by mild solar photo-Fenton in MWTP effluents. Water Reaserch. 2010;**44**:545-554. DOI: 10.1016/j.watres.2009.09.059

[77] Ma D, Yi H, Lai C, Liu X, Huo X, An Z, et al. Chemosphere critical review of advanced oxidation processes in organic wastewater treatment. Chemosphere. 2021;**275**:130104. DOI: 10.1016/j.chemosphere.2021.130104

[78] Pouran SR, Abdul Aziz AR, Wan Daud WMA. Review on the main advances in photo-Fenton oxidation system for recalcitrant wastewaters. Journal of Industrial and Engineering Chemistry. 2015;**21**:53-69. DOI: 10.1016/j.jiec.2014.05.005

[79] Gurung K, Ncibi MC, Shestakova M, Sillanpaa M. Environmental removal of carbamazepine from MBR effluent by electrochemical oxidation. Applied Catalysis B. 2018;**221**:329-338. DOI: 10.1016/j.apcatb.2017.09.017

[80] García-espinoza JD, Mijaylova P. Degradation of pharmaceutical compounds in water by oxygenated electrochemical oxidation: Parametric optimization, kinetic studies and toxicity assessment. Science of the Total Environment. 2019;**691**:417-429. DOI: 10.1016/j.scitotenv.2019.07.118

[81] Geneste F. Catalytic electrochemical pre-treatment for the degradation of persistent organic pollutants. Current Opinion in Electrochemistry. 2018;**11**:19-24. DOI: 10.1016/j.coelec.2018.07.002

[82] Ligneris E. Nanofiber-based materials for persistent organic pollutants in water remediation by adsorption. Applied Sciences. 2018;**8**:166. DOI: 10.3390/app8020166

[83] Liu Y, Hu C, Lo S. Direct and indirect electrochemical oxidation of amine-containing pharmaceuticals using graphite electrodes. Journal of Hazardous Materials. 2019;**366**:592-605. DOI: 10.1016/j.jhazmat.2018.12.037

[84] Kallawar GA, Barai DP, Bhanvase BA. Bismuth titanate based photocatalysts for degradation of persistent organic compounds in wastewater: A comprehensive review on synthesis methods, performance as photocatalyst and challenges. Journal of Cleaner Production. 2021;**318**:128563. DOI: 10.1016/j.jclepro.2021.128563

[85] Kumar A, Anand R, Verma N. Chemosphere efficient electro-oxidation of diclofenac persistent organic pollutant in wastewater using carbon film-supported Cu-rGO electrode. Chemosphere. 2020;**248**:126030. DOI: 10.1016/j.chemosphere.2020.126030

[86] Lan Y, Coetsier C, Causserand C, Groenen-Serrano K. On the role of salts for the treatment of wastewaters containing pharmaceuticals by electrochemical oxidation using a boron doped diamond anode. Electrochimica Acta. 2017;**231**:309-318. DOI: 10.1016/j.electacta.2017.01.160

[87] Wohlmuth da Silva S, Matos do Prado J, Arenhart Heberle AN, Bernardes AM. Electrochemical advanced oxidation of atenolol at Nb/BDD thin film anode. Journal of Electroanalytical Chemistry. 2019;**844**:27-33. DOI: 10.1016/j.jelechem.2019.05.011

Chapter 6

Nonthermal Mechanochemical Destruction of POPs

Giovanni Cagnetta and Mohammadtaghi Vakili

Abstract

The present chapter is dedicated to all relevant theoretical and application aspects of mechanochemical destruction technology for mineralization of POPs, both stockpiled ones and as contaminants in environmental and waste matrices. It will show that such solid-state technology, realized by high energy milling of POPs with a co-milling solid reagent, can achieve complete mineralization of haloorganics into graphitic/amorphous carbon, carbon oxides, and halides; it takes place at near environmental temperature, thus limiting unintentional formation of dioxins (if treatment conditions are selected carefully); and, in some cases, it can be used to produce useful materials instead of just detoxified waste. The chapter will also give a comprehensive picture of complex mechanochemical destruction mechanism, including mechanochemical activation of the co-milling reagent and the cascade of radical reactions that cause POP molecules mineralization. Finally, technological and economic considerations will be provided, which corroborate the validity and feasibility of the mechanochemical destruction as an effective and safe technology to treat POPs.

Keywords: mechanochemistry, high energy milling, POPs mineralization, nonthermal technology, waste detoxification

1. Introduction

As a consequence of their classification as POPs by the Stockholm Convention and prohibition of their use, a number of already manufactured toxic chemicals have become obsolete. Many countries, especially developing ones, do not have the economic and/or technological capacity to dispose such waste materials in proper manner, that is, ensuring their mineralization to non-toxic form. Hence, obsolete chemicals are just stockpiled, often in poor conditions that cannot avert secondary contamination.

Currently, (high temperature) incineration is the sole largely available technology for efficient and economic POPs destruction. However, during combustion notable amounts of hydrogen halide gases are generated, which corrode facility structural elements and, therefore, heighten maintenance cost. Most importantly, risk of unintentional formation of dioxins is never null for such kind of plants. Consequently, new technologies alternative to combustion are highly needed. They must insure complete mineralization of POPs, even at high concentrations, and prevent secondary formation of new POPs [1]. Among the nonthermal alternatives, mechanochemical treatment is considered a valid option for POPs destruction.

1.1 Fundamentals of mechanochemistry

Mechanochemistry is a branch of chemistry that deals with physical and chemical transformations undergone by materials (eminently solids) that are induced during the action of mechanical forces (shear and compression), or are triggered by them [2]. The earliest known example of mechanochemical reaction is friction of two flints that originates sparks. Flint is a variety of quartz, so its scraped surfaces expose radicals that violently react with air producing sparkling plasma. The first mechanochemical experiments are acknowledged to Walthère-Victor Spring (1880–1911), who obtained barium carbonate by solid-state reaction between barium sulfate and sodium carbonate under high pressure; and Mathew Carey Lea (1823–1897), who demonstrated that heating and friction can induce, in some cases, diverse chemical transformations [3]. Today knowledge on mechanochemical phenomena is rather advanced, as well as development of special mechanochemical reactors that can be used to provide high mechanical energy input to solid systems.

A number of physicochemical phenomena may occur to solids under the action of mechanical forces. Some of them are quite unique, like emission of electrons [4] and light [5], while others are commonly experienced such as particle breakage and heating [6]. In truth, a complete classification of the various mechanochemical phenomena is hardly compilable (although some attempts were done [2]). The reason is that often it is not facile to distinguish properly said mechanochemical phenomena (those that occur while the mechanical force acts on the solids) from those that are triggered and/or facilitated by mechanical forces. In addition, it should be noted that every phase transition or emergence of new phases with diverse specific volume (due, for example, to chemical reactions) causes mechanical stress in the solid, thus possibly having mechanochemical effects on the system [7].

Currently, the "mechanochemical activation" theory is considered a reliable explanation of the evolution of solid systems under the action of (sufficiently intense) mechanical forces. Briefly, it hypothesizes that atoms or molecules are shifted from their equilibrium crystal lattice positions by mechanical stress, thus accumulating potential energy (**Figure 1**). This brings the solid in a high-energy metastable state that must release exceeding energy. Common relaxation pathways are heating and particle fracture. But, when particles reach a critical size, solid materials begin to build up crystal defects, develop amorphous phases or other crystalline morphologies, and chemical reactions might take place. Such processes are jointly named "mechanochemical (or mechanical) activation" of solids [8]. Mechanochemically activated solids are more prone to react with other chemicals and can give origin to reactive species.

1.2 Mechanochemical reactors

Action of mechanical forces on solid materials is realized in special mechanochemical reactors. In general, the main effect of these forces is particle breakage and, therefore, the large majority of such machines were originally designed to comminute particles. Consequently, high energy mill is usually considered synonym of mechanochemical reactor. A large number of milling machine typologies is available on the market that can apply mechanical forces with different intensities and by diverse means [9]. Principally, these machines can be classified according to the main type of action in three groups [10]:

1. Shocking mills, where the force derives from high speed impact of the material particles onto elements of the mill. In pin disintegrators, particle hits rotating

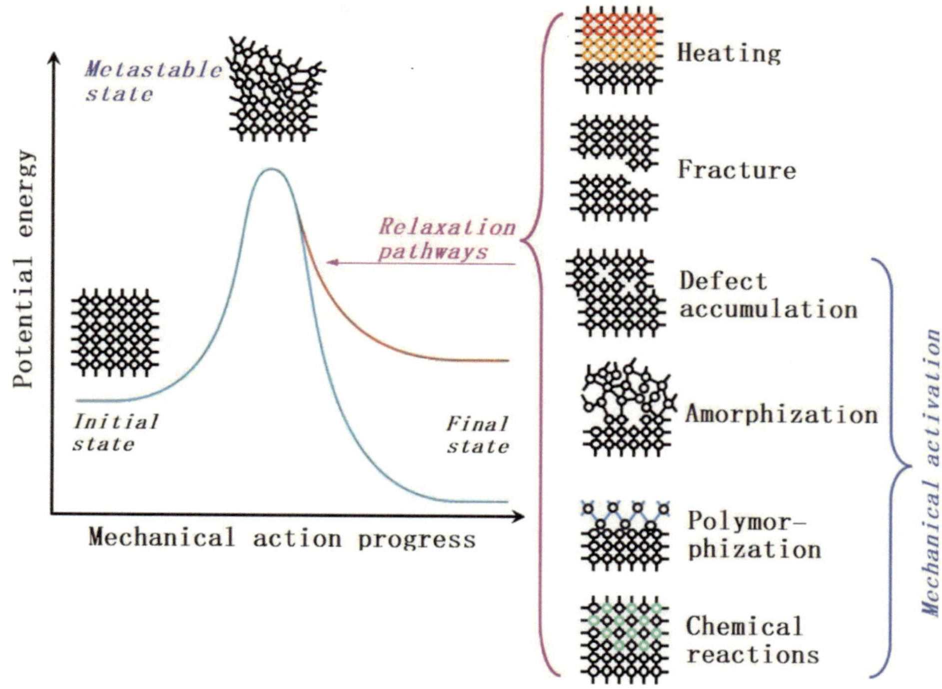

Figure 1.
Main processes occurring during the action of mechanical forces on solids according to the "mechanochemical activation" theory.

blades; in jet mills, a fluid (air) accelerates the particles and throws them against a target.

2. Shear mills, where sliding surfaces of the mill produce friction on the particles, like in ring mills and roller mills.

3. Ball mills, where milling bodies (mostly balls) are accelerated into chambers and provide compression and shear to the material. Such kind of mills can accelerate the milling bodies by planetary, elliptical, etc. movements of the chamber, or by stirring with armed shaft.

Operating parameters of each type of mill mainly control the rate of provision of mechanical energy (i.e., milling intensity) to the milled material, as well as the efficiency of the energy transfer (i.e., the amount of energy that is effectively accumulated by the solid, compared to that dissipated by heat). The quantity of mechanical energy that is accumulated by a unitary mass of milled solid is often referred to as 'specific energy dose'. It has been amply ascertained that transformation degree undergone by a mechanochemical system mainly depends on the total energy dose that is transferred to the system by the high-energy mill, independently from the milling intensity. In other words, the accumulated energy is invariant for a specific mechanochemical system [11]. Indeed, it was also proved that the modality of energy provision, that is, the type of high energy mill has, within certain tolerance ranges, limited influence on the transformation degree [12]. This fact points out that scaling up of mechanochemical processes is relatively facile and is not necessarily done just by trial and error. Taking advantage of the invariance of mechanochemical systems, provision of the same amount of energy dose in high energy mills with diverse scale is a good starting point to achieve similar transformations and transfer the process from small scale to large one.

2. Mechanochemical destruction of POPs

In 1994, Rowlands et al. [13] demonstrated that high energy milling (HEM) can effectively destroy dichlorodiphenyldichloroethane (DDT) in presence of CaO as co-milling reagent. They obtained almost entire dechlorination of DDT, which, as ascertained in subsequent works, was transformed into halides and amorphous/graphitic carbon. Since that groundbreaking work, efficacy of mechanochemical destruction (MCD) obtained by HEM of many toxic organohalogens, included all POPs, has been confirmed [14] (**Figure 2a** shows some examples with CaO as co-milling reagent). In particular, a number of other co-milling reagents and related optimal milling conditions have been investigated, as well as their mechanical activation. Moreover, key aspects of organics mineralization mechanism have been ascertained.

2.1 Treatment conditions of mechanochemical destruction

Co-milling reagent is certainly the most important component of MCD reactions. Theoretically, POPs can be degraded by the sole action of mechanical forces [22], but it would require a long time (i.e., high energy consumption) and could only achieve incomplete mineralization. Differently, co-milling reagent boosts the reaction rate and assures the complete transformation of POPs into inorganic form (usually, halides and carbon). HEM facilitates formation of fresh surfaces on reagent particle due to their breakage, as well as mixing and contact with POPs, thus accelerating the solid-state reaction. Moreover, reagents can be activated by the mechanical energy provided by the mechanochemical reactor, thus enabling or heightening their reactivity (see Section 2.2). Reagents can be classified in four groups:

1. Reducing reagents, like zero valent metals (e.g., Fe, Al, Zn, and Mg) and hydrides (e.g., CaH_2, $NaBH_4$, and $LiAlH_4$).

2. Oxidants, as manganese dioxide (MnO_2), persulfate ($S_2O_8^{2-}$), and ferrate (FeO_4^{2-}).

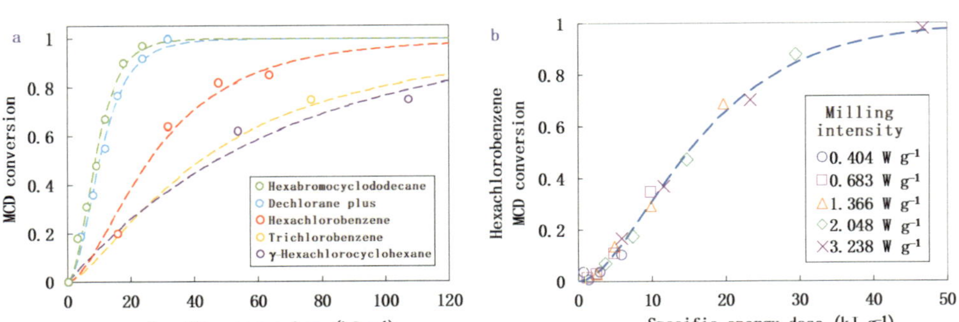

Figure 2.
(a) MCD of some haloorganics (Hexabromocyclododecane [15], Dechlorane plus [16], Hexachlorobenzene [17], Trichlorobenzene [18], and γ-Hexachlorocyclohexane [19]) co-milled with CaO (dashed lines are obtained by interpolation of experimental data with the model of ref. [20]). (b) MCD kinetics invariance respect to milling intensity of hexachlorobenzene co-milled with CaO [21].

3. Lewis bases, such as metal oxides (e.g., CaO and MgO), and strong bases, like NaOH and KOH.

4. Plasma-former reagents, like silica (SiO_2) and alumina (Al_2O_3).

Reagent-to-pollutant ratio (often calculated as mass ratio) is one of the most critical parameters of MCD treatment. It governs the reaction kinetics: within certain ranges, the MCD kinetic constant for a specified POPs-reagent system is directly proportional to such ratio, that is, the higher is the ratio, the faster is the reaction [21]. An exceedingly high reagent ratio, however, decreases the energy efficiency (the amount energy spent to achieve a certain POPs dehalogenation/mineralization percentage) of the treatment, having a negative economic impact.

Milling operation parameters also have a notable influence on the reaction progress. Each HEM device has a number of such parameters that can be modified for the same equipment. In general, some of them are related to geometrical feature of the milling device such as milling chamber dimensions, milling tool (i.e., ball) dimension, and chamber filling ratio; and others are properly said operating parameters, like milling jar speed (e.g., rotation speed or vibration frequency) and ball-to-powder charge ratio. All of them control the amount of the mechanical energy that is inserted in the MCD system and, consequently, its kinetics, but the operating parameters have surely the most relevant effect. It was amply verified that each parameter has an optimal value that maximizes the reaction rate for a specified MCD system [14], which is due to the energetic efficiency of the milling tool impacts. Moreover, it was ascertained as well that MCD systems are invariant respect to the energy provided to the system. In other words, the pollutant degradation conversions are the same for the same amount of mechanical energy inserted, independently from the milling intensity (**Figure 2b**). Hence, MCD results, within certain ranges of tolerance, can be reproduced on any type of HEM device [21].

2.2 Mechanochemical activation of the co-milling reagent

Primary role of HEM is to ensure intimate contact among POPs molecules and the co-milling reagent. In particular, products of the mechanochemical reaction (e.g., carbon), in addition to side-products deriving from interaction with air, milling tools, etc. (e.g., passivating oxide layer on zero valent metals), are removed by the continuous particle fracture, thus exposing fresh surfaces available for further reaction. Such effect is sufficient for reactive materials, like zero valent metals, hydrides, and strong oxidants or bases, which, in some cases, can degrade haloorganics even by simple manual grinding [23].

Other kinds of reagent are efficient thanks to the physicochemical transformations they undergo during HEM. In these cases, the elevated energy input of HEM devices induces the formation of active species, mainly electrons and radicals, that are responsible for POPs mineralization. Metal oxides are known to be very efficacious co-milling reagents. The oxide anion on particle surfaces, thanks to the mechanical energy, generates in the crystal lattice an oxygen vacancy (V_O) with two trapped electrons and an oxygen atom that is released in gaseous form (Eq. (1)). This reaction passes through a step of electron release from the oxide anion to form an oxide radical ($O^{-\bullet}$):

$$O^{2-} \xrightarrow{\Delta E_M} O^{-\bullet} + e^- \xrightarrow{\Delta E_M} V_O + 2\,e^- + \frac{1}{2}O_2\uparrow \qquad (1)$$

Generation of trapped and free electrons in CaO + chlorobyphenyl system, as well as the existence of the oxide radical, was ascertained by electron paramagnetic resonance [24]. Moreover, using the stable radical 2,2-diphenyl-1-picrylhydrazyl (DPPH) as probe, it was verified that electron generation on metal oxide surfaces under HEM is constant and follows a pseudo-zeroth order kinetics [25]. Electrons and oxide radicals are both responsible of haloorganics mineralization, as illustrated in subsection 2.3.

Another class of co-milling reagents that are remarkably reactive under HEM conditions is that of quartz (SiO_2), alumina (Al_2O_3) and silico-aluminates (including clays, etc.). Most of such minerals are known to be plasma-formers, that is, their new surfaces created by particle breakage are rich of electrons [26], which interact with the organic pollutant and mineralize it. For example, in silica-based minerals electrons come from the homolytic cleavage of Si–O bonds, producing silyl ($Si^{\bullet}$) and siloxyl ($SiO^{\bullet}$) radicals [17]. Experiments with DPPH probe suggest that generation kinetics of such species follows the pseudo-first order [27].

Persulfate, a strong oxidant that is being widely utilized for advanced oxidation of organic pollutants in solution, has been proficiently used as co-milling reagent for POPs destruction. As in solution, it is transformed in sulfate radicals ($SO_4^{-\bullet}$) by the mechanical energy, which then oxidize the haloorganics. In presence of strong bases or electron donors, such conversion is faster, thus increasing the overall mineralization rate [28, 29].

2.3 Mechanochemical mineralization of POPs

The active species generated by mechanochemical activation of co-milling reagent, or the reagent itself, interact with POPs molecules triggering and sustaining their degradation, eventually to mineral form. It was recently proposed that the mechanochemical activation of co-milling reagents and the mineralization of haloorganics are kinetically independent, which, in turn, suggests that both processes do not interfere significantly one with the other during HEM [20]. Since reactive species are of two types (i.e., electrons and radicals), their attack generally provokes expulsion of a halide and transformation of the haoloorganics into radical form (**Figure 3**). For example, it was suggested that the first step of sulfonated perfluoroalkyl substances (e.g., perofluorooctane sulfonate) MCD with La_2O_3 is cleavage of the polar group by addition of oxide radical to form perfluorinated moiety radical and sulfate (Eq. (2)) [30]:

$$C_8F_{17} - SO_3^- + O^{-\bullet} \rightarrow \bullet C_8F_{17} + SO_4^{2-} \quad (2)$$

Then, the perfluorinated moiety could undergo further oxidation by reaction with the oxide radical through a so-called CF_2 flaking-off process to form CO_x (or carbonates) and fluorides (Eq. (3)); or it could be reduced by electron addition followed by fluoride expulsion and generation of graphitic/amorphous carbon (Eq. (4)).

$$\cdot C_8F_{17} + 2\,O^{-\bullet} \rightarrow \bullet C_7F_{15} + CO_2 + 2\,F^- \quad (3)$$

$$\cdot C_8F_{17} + 2\,e^- \rightarrow \cdot C_7F_{15} + C + 2\,F^- \quad (4)$$

Likewise, hexachlorobenzene dechlorination in presence of CaO might proceed by capture of an electron from the oxide surface to generate pentachlorobenzyl radical (Eq. (5)) or by substitution of one chlorine with oxide radical to form a pentachlorophenoxyl radical (Eq. (6)) [31]:

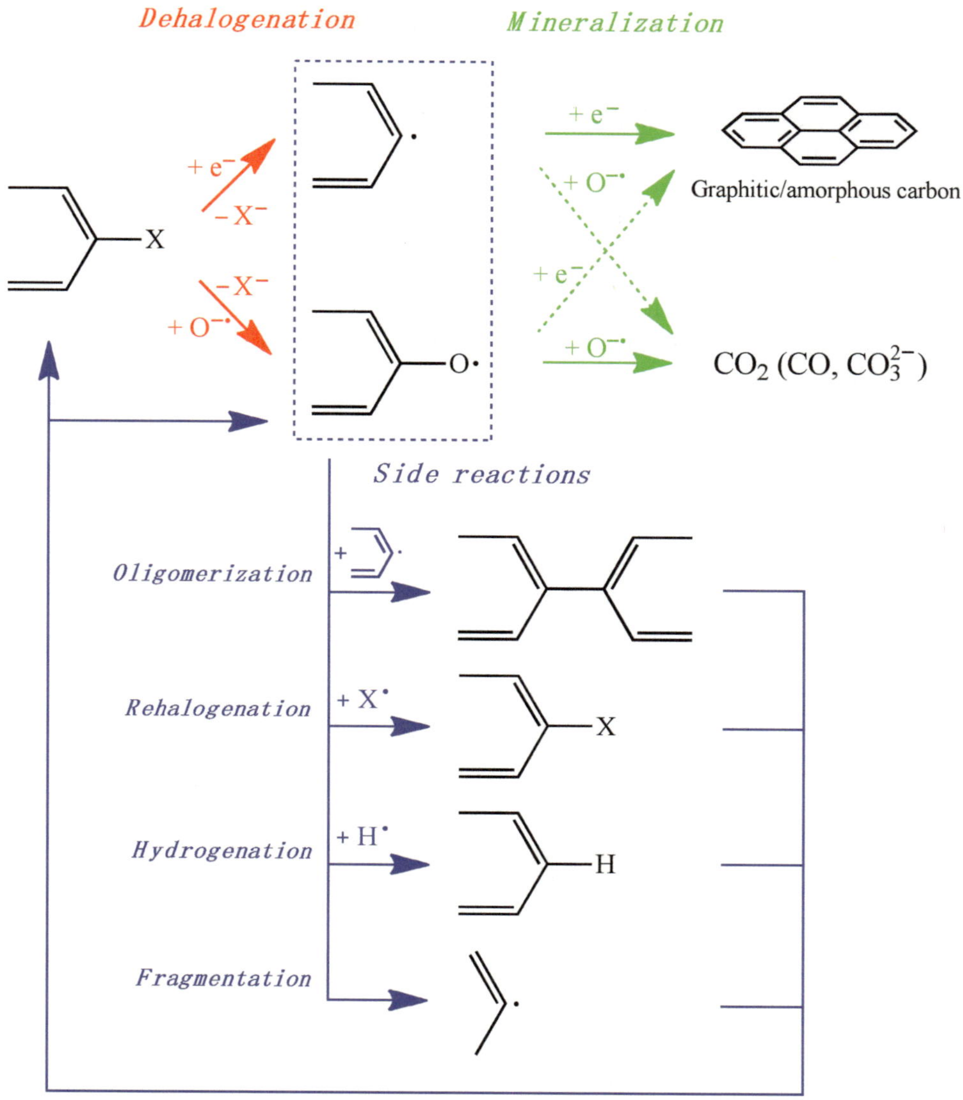

Figure 3.
Generic reaction scheme of MCD process.

$$C_6Cl_6 + e^- \rightarrow \cdot C_6Cl_5 + Cl^- \tag{5}$$

$$C_6Cl_6 + O^{-\cdot} \rightarrow C_6Cl_5O\cdot + Cl^- \tag{6}$$

The mineralization process proceeds in similar fashion by addition of electrons to generate chlorides and graphitic/amorphous carbon, or by oxide radical attack to produce carbon oxides and chlorides.

It can be seen that generation of anion and organic radical as products of active species attack appears to be the general rule for MCD triggering processes. Then, the organic radicals are ultimately transformed into graphitic and amorphous carbon under the attack of electrons, or carbon oxides by addition of oxide radicals. Both redox processes occur at the same time, which is a distinct feature of mechanochemical reactions of organics [32]. Nevertheless, the mineralization process is not that plain (**Figure 3**). A number of secondary radical reactions have been observed in MCD systems, like de−/hydrogenation, oligomerization or radical addition, rehalogenation, etc. They are typical radical reaction that take place among the organoradicals generated during the MCD process. Such by-products, however, are eventually destroyed following the mineralization pathways mentioned above.

Finally, it should be reminded that some types of reagent are per se highly reactive, so they do not necessitate of any mechanochemical activation and the effective mixing realized in the HEM is sufficient to induce the reaction. Obviously, the mineralization depends on the specific reagent. For instance, highly electropositive metals (Na, Mg, etc.) dehalogenate POPs, keeping their organic structure almost intact [33], while less electropositive zero valent metals such as iron produce graphitic/amorphous carbon, probably due to a less effective electron transfer rate, so that the original carbon skeleton is destroyed [17]. Another example is the notable efficacy of KOH to defluorinate perfluoroalkyl substances under HEM. After splitting the polar group, hydroxide anions sequentially substitute fluorides in the perfluorinated moiety, causing shortening of the organic chain by CF_2 flake-off to generate formate [34].

3. Application to stockpiled POPs

Laboratory results on MCD with various reagents can be easily applied to the disposal of stockpiled POPs, which often are a mixture of congeners and/or by-products of the manufacturing process. Anyway, such components have similar reactivity under HEM conditions. Preliminary scaling-up from laboratory-scale to large one can be done by taking advantage of the energetic invariance of mechanochemical reactions (as mentioned in subsection 1.2). Yet, pilot-scale testing could be necessary, especially if the experimental results are translated to a large milling equipment with different type of action, compared to the laboratory one. Choice of the reagent is a vital issue: it should be cheap, easily suppliable, durable, etc., but, its most important feature is efficacy. Efficacious reagent can be utilized with low reagent-to-pollutant ratio (which has a correlation of direct proportionality with MCD rate), so that the energy consumption per mass of treated POPs is contained. Unfortunately, most of the cheapest and largely available co-milling reagents (e.g., CaO, SiO_2, Al_2O_3, Fe, etc.) are not so efficacious and necessitate of large reagent ratios [14]. In order to obviate to this problem, two strategies have been proposed to dispose stockpiled POPs: multi-reagent approach and waste-to-materials one.

3.1 Multi-reagent approach

The multi-reagent approach is simply based on mixing two or more cheap co-milling reagents that, because of their specific physical or chemical properties, have a synergistic interaction that boosts the MCD rate. A typical example of taking advantage of physical properties is the case of mixing a soft reagent (e.g., zero valent metal, metal oxides) with a hard material (e.g., silica, alumina) to improve millability of the former. During HEM, soft material particles reach rapidly the critical size and cannot be further comminuted, therefore their specific surface remains unvaried during the MCD treatment. Moreover, if the reagent is plastic (like metals), the phenomenon of cold-welding hinders particle size reduction and thus the reactivity. Addition of a hard component to the mixture helps an effective fracture of the soft material particles, which are crushed on those of the hard component, forming smaller particles that cover the hard ones. An example is given by the case of zero valent iron (Mohs hardness of 4-5), which had a very poor effectiveness in destroying hexachlorobenzene (**Figure 4a**) [17]. Addition of quartz sand (Mohs hardness of 7), which itself performed better than iron in mineralizing

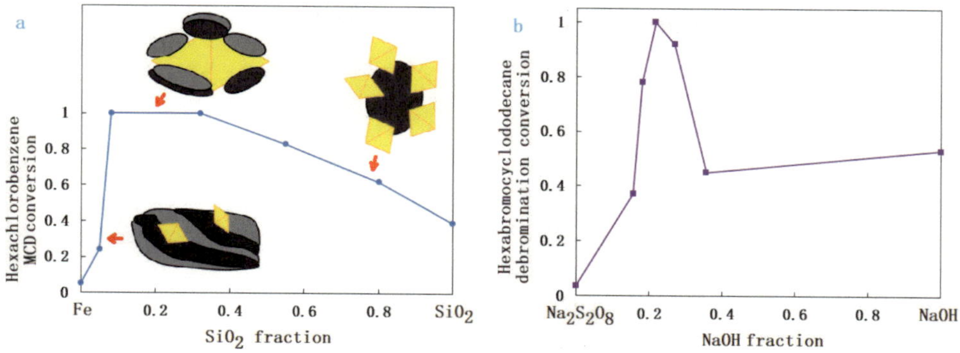

Figure 4.
Influence of multi-reagent composition on two MCD systems: (a) hexachlorobenzene high-energy milled with Fe-SiO_2 [17], and (b) hexabromocyclododecane high-energy milled with $Na_2S_2O_8$-NaOH [35].

the haloorganic, enhanced notably degradation conversion of hexachlorobenzene. Varying the composition of the Fe-SiO_2 mixture, it can be seen that with low silica fractions, such mixture is still scantily effective, since the few SiO_2 crystals are incorporated into the iron cold-welded particles. Then, a range of maximum effectiveness is observed: within this interval, silica crystals are covered with tiny Fe particles, which are extremely reactive towards hexachlorobenzene. Further addition of silica has a negative effect because of redundant number of crystals that cover iron particles.

Additional component(s) can be used to activate or potentiate chemical reactivity of the main co-milling reagent. In this case the synergistic effect depends on the specific chemical properties of the components. For instance, persulfate ($S_2O_8^{2-}$) can be directly activated by sole HEM to generate sulfate radicals ($SO_4^{\cdot-}$) with strong oxidant power. Nevertheless, addition of strong bases or electron donors (e.g., Fe) has been proved to remarkably accelerate persulfate cleavage kinetics and, consequently, target organics mineralization rate. Experiments on hexabromocyclododecane MCD (**Figure 4b**) [35] revealed that the meager debromination capability of persulfate could be markedly improved by addition of ~20% NaOH. Then, excessive NaOH interfered with sulfate generation (likely by reacting with sulfate radical), reducing the debromination rate to levels close to those obtainable by treatment with sole NaOH.

3.2 Waste-to-materials approach

A serious issue of the MCD technology is production of large amounts of HEM residue. In general, the residue is mainly composed of unreacted reagent (since it is often employed in large excess to ensure a rapid and complete destruction of the treated POPs) and mineralization products (graphitic/amorphous carbon and halides). Although detoxified, such material is still an economic burden, because it must be disposed properly. The waste-to-materials approach is aimed to solve such problem by generating a useful material instead of waste. In fact, such method is based on the employment of highly reactive (and rather expensive, too) reagents in stoichiometric amounts that, however, can mineralize POPs and produce a value-added material at the same time.

So far, only two reagents have been ascertained to satisfy both such requirements, that is, bismuth oxide (Bi_2O_3) and lanthanum oxide (La_2O_3). These oxides were used in stoichiometric amount with some brominated and fluorinated POPs (i.e., metal-to-halogen atomic ratio of 1) to mechanosynthetize the corresponding oxyhalide [30, 36, 37]:

$$Br - POPs + Bi_2O_3 \rightarrow BiOBr + C + BiCO_3Br\ (+CO_2) \quad (7)$$

$$Br - POPs + La_2O_3 \rightarrow LaOBr + C + LaCO_3Br\ (+CO_2) \quad (8)$$

$$F - POPs + La_2O_3 \rightarrow LaOF + C + LaCO_3F\ (+CO_2) \quad (9)$$

POPs were entirely mineralized in graphitic/amorphous carbon and CO_2 (mainly found as carbonate), thus ensuring detoxification. At the same time, an almost pure oxyhalide was obtained, after a short thermal treatment to remove C and reconvert the carbonate into the corresponding oxyhalide. Bismuth oxybromide is a material with excellent photocatalytic properties and the mechanosynthetized BiOBr was tested for removal of dye methyl orange in water under visible light irradiation. Lanthanum oxyhalides have excellent optical properties with actual application in X-ray imaging for medical devices (LaOBr) and potential one to produce doping host for transparent oxy-fluoride glass ceramics (LaOF). The production of such value-added materials could be a driving force to use toxic and obsolete POPs as source of halogens, achieving their detoxification.

4. Application to contaminated waste

MCD of POPs in contaminated waste is complicated by components of the waste matrix, making it almost unpredictable. Most of the components are mechanochemically activated by HEM, so they can interact with both POPs and co-milling reagent(s). Such interaction could be positive or negative. Components such as aluminosilicates, metal oxides, carbonates, etc. can be more or less activated, thus supporting the mineralization process. Moreover, some of these components are known to acquire improved catalytic properties during and after HEM, frequently facilitating POPs degradation [38]. On contrary, radical scavengers, like organic matter, hinder the degradation of haloorganics. In the following subsections, three examples are discussed: soils and sediments, fly ashes, and plastic waste.

4.1 Soils and sediments

In general, soils and sediments are suitable matrices to obtain effective mineralization of POPs by MCD. Providing an adequate amount of mechanical energy (e.g., prolonging HEM for sufficiently long time) ensures destruction of haloorganics due to a number of phenomena that might occur during the treatment [39]. In the first place, decrease of particle size and the consequent enlargement of specific surface enhance adsorption capacity of soils and sediments towards POPs. Clays play a key role in this: aside from particle breakage, HEM induces delamination of aluminosilicates, as well as partial amorphization of their surfaces, thus exposing more dangling bonds [40]. Hence, POPs can be adsorbed mainly by Van der Waals interactions, and possibly undergo catalytic degradation, thanks to surface acidity of clays [38].

More relevantly, aluminosilicates, metal oxides, carbonates, and other inorganic components of soil and sediments can be mechanochemically activated to generate active species (as elucidated in Section 2.2). These are deemed to be the major responsible of POPs mineralization in such kinds of waste [27, 39]. Organic matter, on the other hand, might facilitate adsorption and catalytic degradation of POPs onto particles, but surely scavenges radical species generated by the mechanical activation of inorganic components and co-milling reagent(s).

Since HEM can activate several mineral components of soil and sediments, MCD can be potentially realized by taking advantage of the "self-healing" properties of these matrices. Indeed, mechanochemical treatment of such contaminated waste without addition of co-milling reagent suffices, in some cases, to achieve entire degradation of POPs [41]. Using a reagent, typically in large amount, insures complete mineralization of the haloorganics in a reasonable time and sensibly boosts the MCD rate [42], but usually transforms the contaminated matrix in a useless waste (whose amount is in general conspicuous). In order to avoid excessive usage of reagent, but to keep energy consumption contained, coupling of MCD with other technologies, such as thermal desorption [43] and biological treatment [44] was also experimented. Such approach is probably the most promising to avert extreme denaturation of the contaminated soils and sediments and allow their relocation in the original geological position.

4.2 Fly ashes

Fly ashes, with their high content of silicates, is another matrix that responds well to MCD treatment. However, because of the notable concentration of PCDD/Fs, such waste must be co-milled with a suitable amount of reagent to ensure entire mineralization of dioxins. Metal oxides (e.g., CaO, MgO), zero valent metals (e.g., Al), and their combination are inexpensive reagents that can efficaciously destroy PCDD/Fs, dioxin-like compounds, and their precursors [45, 46].

A key issue of fly ash detoxification by MCD is the compresence of carbon and chlorides, which determines a high potential for reformation of dioxins, also under low-temperature HEM conditions. In fact, de novo formation of PCDD/Fs was observed during the mechanochemical treatment [46, 47], in particular in presence of dioxin-formation catalysts, like copper compounds [45]. This might be caused by hits of the milling tools, which induce high local temperature increase, although for short time, on surfaces of the particle that are trapped between the hitting tools [45]. Despite such issue, it was proved that sufficiently long time milling, as well as temperature control, assures definitive dioxin removal and prevents their reformation [48]. This is owed to extensive amorphization of the fly ash components, especially carbonaceous matter, which averts de novo formation [45, 47].

4.3 Plastic waste

Some types of plastic waste contain high amounts of (brominated) flame retardants because of their utilization in electric and electronic devices. Removal of such chemicals from the polymeric matrix is a hard task. Nonetheless, it was realized by HEM with co-milling reagents such as zero valent metals (e.g., iron), metal oxides (e.g., CaO), plasma-formers (e.g., SiO_2), and their combination [49, 50]. Reagents with relatively high hardness (i.e., Fe and SiO_2) were found to be more efficacious to debrominate the plastic waste for the reason that they improve the mechanical and chemical degradation of the polymeric matrix, thus allowing a better contact between the reagent and the flame retardant.

Another relevant problem of brominated flame retardant MCD in plastic waste is the negative effect of the polymeric matrix, which slows down the degradation rate. Firstly, the impact energy is mainly absorbed by the matrix, and only a minor share is actually available for reagent activation and debromination of the POPs. Secondly, the mechanochemically activated radical species generated from the co-milling reagent are scavenged by the polymer, leading to chain shortening and other degradation phenomena of the plastic. In fact, it was verified experimentally that decabromodiphenyl ether mechanochemical degradation rate in polypropylene matrix co-milled with Fe-SiO_2 mixed reagent was 4.4 times slower than the rate observed for the pure flame retardant co-milled with the same reagent and under similar HEM conditions [49]. Consequently, longer milling times and higher energy consumption are required for POPs mineralization in plastic waste, compared to the MCD of sole haloorganics.

5. Technological and economic considerations

The MCD technology can efficaciously destroy POPs, whether they are in almost pure form as (obsolete) chemicals, or they are present in environmental and waste matrices as contaminants. In both cases it is possible to transform the POPs in mineral form using HEM, often employing a co-milling reagent. Yet, effectiveness alone is not sufficient for large-scale application of this technology: greenness, safeness, and cost effectiveness are necessary requirements as well. Among non-thermal technologies, MCD is certainly one of the greenest, compared to other POPs destruction technologies. It does not require any solvent, since it is a solid-state treatment. And, it can be potentially utilized to prepare useful materials (instead of detoxified waste). More importantly, MCD phenomena occur only under the energy input provided by the HEM and can be terminated by simply turning off the milling device. Hence, in case of any non-mechanical accident (e.g., unintentional emission of toxic chemicals), the process can be interrupted immediately [37]. In addition, when treating hazardous waste with high potential of dioxin de novo formation, milling chamber temperature can be kept very close to environmental temperature to avert unintentional dioxin generation [48]. In sum, MCD is a safe and green technology.

Another substantial advantage of the MCD technology is the simplicity of the plant design and its versatility. **Figure 5** shows a block scheme of a mechanochemical plant for treatment of stockpiled POPs and POPs-contaminated waste. Stockpiled POPs are directly fed into the milling section, with a co-milling reagent. An air treatment section is included to prevent any emission due to volatilization of the POPs or their degradation by-products, as well as release of contaminated dust. MCD treatment of waste materials just includes (depending on the specific waste to be treated) a drying section to decrease humidity content, and a sieving section to remove debris that cannot be fed to the mill (e.g., stones in contaminated soil). Air deriving from each of these auxiliary sections is treated to avert possible POPs release in the environment. Obviously, such simple plant scheme can be used versatilely to treat any type of POPs waste.

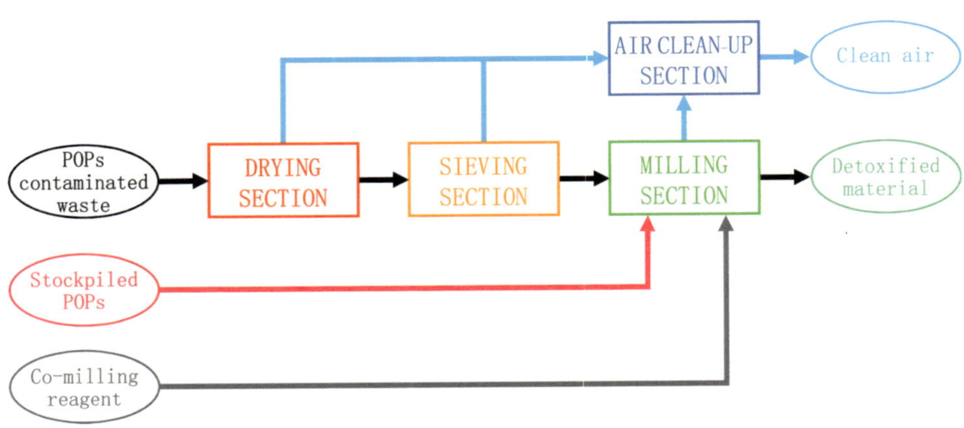

Figure 5.
Block scheme of a mechanochemical plant for stockpiled POPs and POPs contaminated waste treatment.

Despite the above-mentioned advantages, MCD technology is affected by two issues: noise and fine powder emission, which is related to worker and environmental safety, and high energy consumption, which is mainly an economic problem. The first one is easily overcome by constructing adequate containment facilities and utilizing individual protective devices inside such facilities. The issue of energy consumption can be managed by a few ways, which can be selected through an adequate economic assessment. Possible options comprise employment of large amounts of cheap and easily suppliable reagents to boost the MCD reaction rate, or, alternately, reduced quantities of strong/efficacious reagents; and coupling MCD with other non- or low-thermal technologies (e.g., biological treatment, thermal desorption, etc.).

Economic assessment is the sole way to evaluate effectively the various solutions for reduction of energy consumption, as well as other issues related to plant configuration. Typology of HEM device available on the market, electric energy cost, kind of the accessible reagent(s), and nature and concentration of the POPs waste are some of the major factors that have remarkable influence on the investment and operating cost of an MCD plant. Such factors depend more or less on the location, so it is not possible to execute a priori a generic economic assessment for this technology. But, a tentative economic feasibility study for MCD treatment for soil was carried out on the basis of data related to the US in 2016 [39]. This study highlighted that milling chamber volume is the key parameter that governs both investment and operating costs: the larger is the volume, the lower are the expenditures. Estimated operating costs were close to or less than those of the technologies currently available on the market for contaminated soil treatment. Actually, the chief economic issue of MCD technology is that most of HEM devices are relatively small, hence, it is for now more suitable to treat low volume waste, like stockpiled POPs. On the other hand, given the increasing interest in mechanochemical technology in various fields, it is expected that larger scale mills will be available on the market, so that MCD treatment of large volume waste such as contaminated matrices will become economical, too.

6. Conclusions

The present chapter illustrated that high energy ball milling of haloorganics in presence of co-milling reagent(s) can dismantle the structure of such organics to generate inorganic graphitic/amorphous carbon, carbon oxides, and halides. This transformation occurs at near-room temperature and pressure; therefore, the chance of unintentional formation of dioxins and dioxin-like compounds is almost null (if milling parameters are chosen judiciously). POPs can be efficaciously destroyed as both in almost pure form (e.g., stockpiled obsolete chemicals) and in contaminated matrices (e.g., soil, sediments, and hazardous waste). The high energy consumption and the lack of sufficiently large industrial high energy milling equipment hamper full-scale application of this technology. Nevertheless, specific approaches to reduce energy consumption, such as multireagent and waste-to-materials strategies, and the increasing interest for mechanochemical methods in other fields, which is pushing also the development of low-priced large-scale mills, will facilitate application of mechanochemical treatment to POPs destruction.

Conflict of interest

The authors declare no conflict of interest.

Author details

Giovanni Cagnetta[1*] and Mohammadtaghi Vakili[2]

1 State Key Joint Laboratory of Environment Simulation and Pollution Control (SKJLESPC), Beijing Key Laboratory for Emerging Organic Contaminants Control (BKLEOC), School of Environment, Tsinghua University, Beijing, China

2 Green Intelligence Environmental School, Yangtze Normal University, Chongqing, China

*Address all correspondence to: gcagnetta@mail.tsinghua.edu.cn

References

[1] USEPA. Reference Guide to Non-combustion Technologies for Remediation of Persistent Organic Pollutants in Soil, Second Edition – 2010. Available from: https://www.epa.gov/sites/production/files/2015-08/documents/pops_report_finalepa_sept2010.pdf [Accessed: 15 Nov 2021]

[2] Boldyreva E. Mechanochemistry of inorganic and organic systems: What is similar, what is different? Chemical Society Reviews. 2013;**42**(18):7719

[3] Takacs L. The historical development of mechanochemistry. Chemical Society Reviews. 2013;**42**(18):7649

[4] Boldyrev V. Experimental methods in the mechanochemistry of inorganic solids. Treatise on Materials Science & Technology. 1983;**19**:185-223

[5] Xie Y, Li Z. The development of mechanoluminescence from organic compounds: Breakthrough and deep insight. Materials Chemistry Frontiers. 2020;**4**(2):317-331

[6] Baláž P, Achimovičová M, Baláž M, Billik P, Cherkezova-Zheleva Z, Criado JM, et al. Hallmarks of mechanochemistry: From nanoparticles to technology. Chemical Society Reviews. 2013;**42**(18):7571-7637

[7] Boldyrev VV. Mechanochemical processes with the reaction-induced mechanical activation. Chemo-mechanochemical effect. Russian Chemical Bulletin. 2018;**67**(6):933-948

[8] Boldyrev VV. Mechanochemistry and mechanical activation of solids. Russian Chemical Reviews. 2006;**75**(3):177-189

[9] Lomovskiy I, Bychkov A, Lomovsky O, Skripkina T. Mechanochemical and size reduction machines for biorefining. Molecules. 2020;**25**(22):5345

[10] Boldyrev VV, Tkačova K. Mechanochemistry of solids: Past, present, and prospects. Journal of Materials Synthesis and Processing. 2000;**8**(3/4):121-132

[11] Delogu F, Deidda C, Mulas G, Schiffini L, Cocco G. A quantitative approach to mechanochemical processes. Journal of Materials Science. 2004;**39**(16/17):5121-5124

[12] Santhanam PR, Dreizin EL. Predicting conditions for scaled-up manufacturing of materials prepared by ball milling. Powder Technology. 2012;**221**:403-411

[13] Rowlands SA, Hall AK, McCormick PG, Street R, Hart RJ, Ebell GF, et al. Destruction of toxic materials. Nature. 1994;**367**(6460):223

[14] Cagnetta G, Robertson J, Huang J, Zhang K, Yu G. Mechanochemical destruction of halogenated organic pollutants: A critical review. Journal of Hazardous Materials. 2016;**313**:85-102

[15] Zhang K, Huang J, Wang H, Liu K, Yu G, Deng S, et al. Mechanochemical degradation of hexabromocyclododecane and approaches for the remediation of its contaminated soil. Chemosphere. 2014;**116**:40-45

[16] Zhang W, Huang J, Yu G, Deng S, Zhu W. Mechanochemical destruction of dechlorane plus with calcium oxide. Chemosphere. 2010;**81**(3):345-350

[17] Zhang W, Wang H, Jun H, Yu M, Wang F, Zhou L, et al. Acceleration and mechanistic studies of the mechanochemical dechlorination of HCB with iron powder and quartz sand. Chemical Engineering Journal. 2014;**239**:185-191

[18] Tanaka Y, Zhang Q, Saito F. Mechanochemical dechlorination of

chlorinated compounds. Journal of Materials Science. 2004;**39**(16/17): 5497-5501

[19] Nomura Y, Fujiwara K, Terada A, Nakai S, Hosomi M. Mechanochemical degradation of γ-hexachlorocyclohexane by a planetary ball mill in the presence of CaO. Chemosphere. 2012;**86**(3):228-234

[20] Vakili M, Qin R, Cagnetta G, Huang J, Wang B, Yu G. Improved fractal kinetic model to predict mechanochemical destruction rate of organic pollutants. Chemosphere. 2021; **284**:131307

[21] Cagnetta G, Huang J, Wang B, Deng S, Yu G. A comprehensive kinetic model for mechanochemical destruction of persistent organic pollutants. Chemical Engineering Journal. 2016; **291**:30-38

[22] Dubinskaya AM. Transformations of organic compounds under the action of mechanical stress. Russian Chemical Reviews. 1999;**68**(8):637-652

[23] Pizzigallo MDR, Napola A, Spagnuolo M, Ruggiero P. Mechanochemical removal of organo-chlorinated compounds by inorganic components of soil. Chemosphere. 2004;**55**(11):1485-1492

[24] Ikoma T, Zhang Q, Saito F, Akiyama K, Tero-Kubota S, Kato T. Radicals in the mechanochemical dechlorination of hazardous organochlorine compounds using CaO nanoparticles. Bulletin of the Chemical Society of Japan. 2001;**74**(12):2303-2309

[25] Cagnetta G, Huang J, Lu M, Wang B, Wang Y, Deng S, et al. Defect engineered oxides for enhanced mechanochemical destruction of halogenated organic pollutants. Chemosphere. 2017;**184**:879-883

[26] Kaupp G. Mechanochemistry: The varied applications of mechanical bond-breaking. CrystEngComm. 2009; **11**(3):388-403

[27] Hu A, Cagnetta G, Huang J, Yu G. Mechanochemical enhancement of the natural attenuation capacity of soils using two organophosphate biocides as models. Journal of Hazardous Materials. 2018;**360**:71-81

[28] Liu X, Zhang X, Zhang K, Qi C. Sodium persulfate-assisted mechanochemical degradation of tetrabromobisphenol A: Efficacy, products and pathway. Chemosphere. 2016;**150**:551-558

[29] Wang N, Lv H, Zhou Y, Zhu L, Hu Y, Majima T, et al. Complete defluorination and mineralization of perfluorooctanoic acid by a mechanochemical method using alumina and persulfate. Environmental Science & Technology. 2019;**53**(14): 8302-8313

[30] Cagnetta G, Zhang Q, Huang J, Lu M, Wang B, Wang Y, et al. Mechanochemical destruction of perfluorinated pollutants and mechanosynthesis of lanthanum oxyfluoride: A waste-to-materials process. Chemical Engineering Journal. 2017;**316**:1078-1090

[31] Deng S, Feng N, Kang S, Zhu J, Yu B, Chen J, et al. Mechanochemical formation of chlorinated phenoxy radicals and their roles in the remediation of hexachlorobenzene contaminated soil. Journal of Hazardous Materials. 2018;**352**:172-181

[32] Leitch JA, Browne DL. Mechanoredox chemistry as an emerging strategy in synthesis. Chemistry – A European Journal. 2021; **27**(38):9721-9726

[33] Birke V, Mattik J, Runne D. Mechanochemical reductive dehalogenation of hazardous polyhalogenated contaminants. Journal

of Materials Science. 2004;**39**(16/17): 5111-5116

[34] Zhang K, Cao Z, Huang J, Deng S, Wang B, Yu G. Mechanochemical destruction of Chinese PFOS alternative F-53B. Chemical Engineering Journal. 2016;**286**:387-393

[35] Yan X, Liu X, Qi C, Lin C, Li P, Wang H. Disposal of hexabromocyclododecane (HBCD) by grinding assisted with sodium persulfate. RSC Advances. 2017;7(38): 23313-23318

[36] Zhang K, Huang J, Wang H, Yu G, Wang B, Deng S, et al. Mechanochemical destruction of decabromodiphenyl ether into visible light photocatalyst BiOBr. RSC Advances. 2014;**4**(28): 14719-14724

[37] Cagnetta G, Liu H, Zhang K, Huang J, Wang B, Deng S, et al. Mechanochemical conversion of brominated POPs into useful oxybromides: A greener approach. Scientific Reports. 2016;**6**(1):28394

[38] Nasser A, Mingelgrin U. Mechanochemistry: A review of surface reactions and environmental applications. Applied Clay Science. 2012;**67–68**:141-150

[39] Cagnetta G, Huang J, Yu G. A mini-review on mechanochemical treatment of contaminated soil: From laboratory to large-scale. Critical Reviews in Environmental Science and Technology. 2018;**48**(7–9):723-771

[40] Ancona V, Di Leo P, Pizzigallo MDR. Sequestration of catechol and pentachlorophenol by mechanochemically treated kaolinite. Clays and Clay Minerals. 2016;**64**(5): 513-522

[41] Wang H, Hwang J, Huang J, Xu Y, Yu G, Li W, et al. Mechanochemical remediation of PCB contaminated soil. Chemosphere. 2017;**168**:333-340

[42] Zhang X, Liu X, Zhao J, Sun W, Li X, Li J, et al. NaOH-activated persulfate-assisted mechanochemical mechanism and removal of lindane from contaminated soil. Journal of Environmental Chemical Engineering. 2021;**9**(4):105391

[43] Zhao Z, Ni M, Li X, Buekens A, Yan J. Combined mechanochemical and thermal treatment of PCBs contaminated soil. RSC Advances. 2017; 7(34):21180-21186

[44] Cagnetta G, Intini G, Liberti L, Boldyrev VV, Lomovskiy OI. The biomec process for mechanochemically assisted biodegradation of PCBs in marine sediments. Journal of Soils and Sediments. 2015;**15**(1):240-248

[45] Cagnetta G, Hassan MM, Huang J, Yu G, Weber R. Dioxins reformation and destruction in secondary copper smelting fly ash under ball milling. Scientific Reports. 2016;**6**(1):22925

[46] Chen Z, Mao Q, Lu S, Buekens A, Xu S, Wang X, et al. Dioxins degradation and reformation during mechanochemical treatment. Chemosphere. 2017;**180**:130-140

[47] Peng Y, Buekens A, Tang M, Lu S. Mechanochemical treatment of fly ash and de novo testing of milled fly ash. Environmental Science and Pollution Research. 2018;**25**(19):19092-19100

[48] Lu M, Lv T, Li Y, Peng Z, Cagnetta G, Sheng S, et al. Formation of brominated and chlorinated dioxins and its prevention during a pilot test of mechanochemical treatment of PCB and PBDE contaminated soil. Environmental Science and Pollution Research. 2017; **24**(24):20072-20081

[49] Cagnetta G, Zhang K, Zhang Q, Huang J, Yu G. Mechanochemical pre-

treatment for viable recycling of plastic waste containing haloorganics. Waste Management. 2018;**75**:181-186

[50] Qin B, Lin M, Chen X, Xu Z, Fu Y, Hu J, et al. A novel approach for determining the accurate debromination time in the ball-milling process of nonmetallic particles from waste printed circuit boards by computation. Journal of Hazardous Materials. 2021; **410**:124611

Chapter 7

Physiochemical Properties and Removal Methods of Phenolic Compounds from Waste Waters

Yesim Gucbilmez

Abstract

In this chapter, phenol and chlorophenols are investigated in terms of their production histories, physiochemical properties, pollution resources, and removal methods. It is seen that both phenol and chlorophenols are highly toxic compounds, produced from natural and anthropogenic sources, which are hazardous to both humans and the environment even at very low concentrations. The typical industries which produce phenol and chlorophenol pollution are petrochemical, textile, plastics, resin, dye, pharmaceutical, iron and steel, pulp and paper industries as well as the petroleum refineries, and coal gasification operations. Phenol is a highly corrosive and nerve poisoning agent. It causes harmful health effects, such as sour mouth, diarrhea, and impaired vision. It is also toxic for the ecosystem with toxicity levels ranging between 10-24 mg/L for humans, 9-25 mg/l for fish, and lethal blood concentration around 150-mg/100 ml. Chlorophenols found in natural waters or drinking water also cause serious health problems such as histopathological alterations, genotoxicity, mutagenicity, and carcinogenicity among others. Due to the aforementioned reasons, the phenolic compounds in wastewaters or drinking water must be removed using a suitable wastewater treatment method such as adsorption, extraction, electrochemical oxidation, biodegradation, catalytic wet air oxidation, or enzyme treatment among others.

Keywords: phenol, chlorophenols, wastewater treatment, phenolic compounds, phenolics, organic pollutants

1. Introduction

Due to technological advances and rapid industrial growth, water systems around the world are under threat. In general, developing countries suffer from water pollution originating from agricultural sources whereas developed countries have chemical discharge problems. In most wastewaters and drinking water, one or more of the following toxic organic pollutants may occur [1]: Organochlorines, chlorobenzenes (CBs), polychlorinated dibenzo-p-dioxins (PCDDs) and polychlorinated dibenzofurans (PCDFs), phenols, chlorophenols, and other phenol-derived compounds.

Phenolic compounds are phytochemicals found in nature that cannot be synthesized in the human body. They are mainly obtained from food and medicinal herbs and are present in most fruits and vegetables [2]. They are found in water systems due to the discharge streams of industrial, agricultural, and domestic activities as well as the result of natural phenomena. They cause severe and long-lasting health hazards including damage to red blood cells and the human liver. Their interaction with the aquatic ecosystem can produce new compounds, which can be as toxic as the original phenolic molecules [3].

Chlorophenols (CPs) are common contaminants that can be found in surface, ground, and drinking waters [4–8]. Thermal and chemical decomposition of chlorophenols leads to the formation of harmful compounds which cause public health problems such as genotoxicity, mutagenicity, and carcinogenicity among others. In addition, some electrophilic molecules may occur as a result of the transformation of chlorophenols, which may connect and harm the DNA or gene products [5, 6].

2. History of phenol and chlorophenols

The academic and industrial chemist F.F. Runge was born in 1794 in Billwerder near Hamburg. Runge isolated phenol from coal tar in the year 1834 in an impure form and gave it the name "carbolic acid" "Karbols¨aure" (carbolic acid) [7–9].

British surgeon Joseph Lister used phenol as a disinfectant for the first time in 1865 for sterilizing surgical dressings, instruments, and wounds [10]. However, the phenol sprays used during the surgeries were dangerous for the lung's mucous membrane when inhaled. Thus, by 1890, the phenol spray was abandoned by the medical community [11].

During the Boer War, England placed an embargo on phenol causing phenol shortage on the continent. This led F. Raschig works at Ludwigshafen, Germany, to produce synthetic phenol on a large scale and in 1940, Hooker Chemical Corporation built a plant based on the so called the Raschig-Hooker process for the commercial production of phenol [12, 13].

During World War I and World War II, the existing sulfonation process for the production of phenol was further improved and other processes such as chlorination process and the Raschig Process were commercialized for the first time [14]. In 1924, Dow Chemical started the commercial production of synthetic phenol using direct chlorination of benzene to chlorobenzene which was called the Dow Process [11].

In terms of patent literature, since the middle of the twentieth century, researchers have mainly tried to improve the existing processes rather than developing new ones. One such example was the 1989 European patent of Mitsui Petrochemical Limited which used a recycle loop to reduce the amount of the side product acetone [15, 16].

Presently, phenol is used in the manufacture of phenolic and epoxy resins [17, 18], plastics [19], plasticizers [20], polycarbonates [21], nylon [22], dyes [23, 24] disinfectants [25], herbicides [26], polymers, drugs, pesticides [27] wood preservatives [28], and fungicides [29].

Chlorophenols (CPs) are produced by the electrophilic halogenation of phenol with chlorine. There are five basic types and 19 different CPs. CPs are mostly used as pesticides, herbicides, antiseptics, and disinfectants [30]. They are very toxic and their presence is dangerous for humans as well as aquatic life. They are found in the wastewaters of textile, pharmaceutical, petrochemical, pesticide, paper, and other industries [31].

The processes given in **Figure 1** have been mainly used in the industry for the production of phenol up to date [10, 32–35] and among them, only the Hock and Toluene Oxidation Processes are important for the phenol industry, the others have been discarded for economic reasons [10].

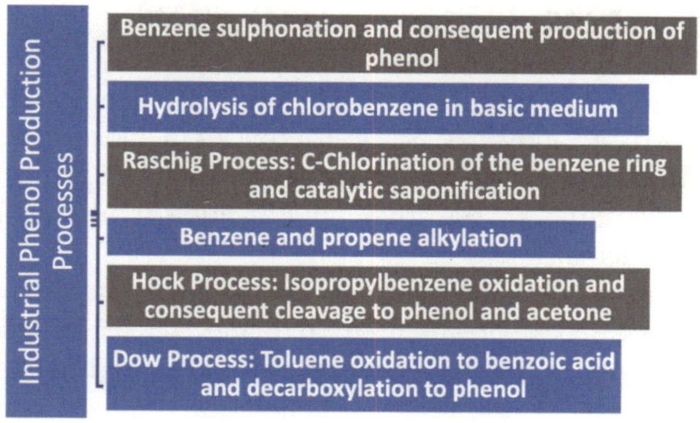

Figure 1.
Processes used up to date for the industrial production of phenol [10, 32–35].

The global phenol market reached a value of 23.17 billion US Dollars (USD) in the year 2020. The phenol market is further expected to grow at a compound annual growth rate (CAGR) of 5.3% between the years 2021 and 2026 to reach a value of about 30 billion USD by the year 2026 [36].

In the case of CPs, their production and usage have caused the presence of persistent toxic components into the water systems which are resistant to biological degradation. However, microorganisms exposed to these pollutants have obtained the ability to biologically degrade some of them and the biodegradation routes were seen to depend on the physicochemical and biological properties of the particular wastewater system under question. Understanding the genetic basis of catabolism of CPs may increase the efficiency of naturally occurring microorganisms or help to raise new microorganisms which can degrade CPs successfully [37].

3. Physiochemical properties of phenol and chlorophenols

Phenol is an odorous chemical compound, found either as a colorless liquid or white solid at room temperature, and maybe highly toxic and corrosive. Phenols are similar to alcohols but form stronger hydrogen bonds. The presence of stronger hydrogen bonds makes them more water-soluble than alcohols and their boiling points higher than those of alcohols [38]. They are widely used as raw materials in the manufacture of phenolic resins, automotive parts, nylon, epoxy resins, polycarbonate engineering thermoplastics, wood preservatives, heavy-duty surfactants, pharmaceuticals, disinfectants, tank linings, and coating materials among others. They are also widely used in household products; for instance, phenol-derived *n*-hexylresorcinol, is used in cough drops and other antiseptic applications and butylated hydroxytoluene (BHT) is a common antioxidant in foods. Also, in the dye industry, substituted phenols are used to make intensely colored azo dyes [38, 39].

CPs, produced by chlorinating phenol or hydrolyzing chlorobenzenes, contain the benzene ring, the OH group, and chlorine atoms. All CPs, except 2-CP, are solids with melting points in the range of 33–191°C. They are weakly acidic and their acidity is slightly lower than that of phenols. In reactions with alkaline metals in water solutions, they yield highly soluble metal salts. Their level of toxicity depends on the chlorination degree and the place of the chlorine atoms with respect to the hydroxyl group [40].

4. Sources of phenol and chlorophenol pollution

The presence of phenolic compounds in wastewaters stem from two main sources: Natural and human-based activities. Natural activities include the decomposition of dead plants and animals, synthesis by microorganisms and plants in the aquatic ecosystem. Human activities, on the other hand, include industrial, domestic, agricultural, and municipal activities [10].

Phenol enters water systems in effluents from major industries, such as petrochemical, textile, plastics, resin manufacturing, dye, pharmaceutical, iron and steel, pulp and paper as well as petroleum refineries, and coal gasification operations. It is very important to remove phenols and aromatic compounds from industrial streams before discharging them because of their toxicity to aquatic organisms [41, 42].

The CPs are among the most important environmental pollutants. They are mostly used in the production of paper and pesticides and also as intermediate materials in the production of dyes, plastics, and pharmaceuticals [43–45]. These industries often cause wastewater and groundwater pollution. In addition, as a result of tap water chlorination treatment, CPs have also been detected in drinking water [46].

5. Removal methods of phenol and chlorophenols

Phenol is a nerve poisoning agent and it is highly corrosive. It causes health hazards, such as diarrhea, sour mouth, and impaired vision. It is also toxic for fish and the toxicity levels are in the range of 10–24 mg/L for humans and 9–25 mg/L for fish while the lethal blood concentration is around 150-mg/100 ml [47].

The available removal methods used for phenol can be separated into two main groups: Traditional and advanced. Traditional methods include steam distillation, extraction, adsorption, ion exchange [42] and advanced methods include wet air oxidation, catalytic wet air oxidation, ozonation, membrane processes, electrochemical oxidation, biological processes/biodegradation, and enzymatic treatment among others [42, 48].

Using the relative volatility of phenol, steam distillation can be carried out in order to remove phenol from aqueous mixtures. The phenol–water mixture forms a minimum azeotrope at 9.21% (w/w) phenol [42, 49–51]. Using this property, azeotropic distillation or steam distillation can be used to treat phenol wastewaters to obtain effluent concentrations as low as 0.01 mg/L [48].

Liquid–liquid extraction is a commonly employed technique used for the removal of phenolic compounds. It can be used for a wide range of phenol concentrations and is economical in some cases. Benzene and butyl acetate have been popular as solvents in this process in the past, however, presently, the most used solvent is di-isopropyl ether, which is used in the phenosolvan process [50, 51]. For the extraction process, the solubility of the preferred solvent in water should be tolerable so that further purification steps will not be necessary. The selectivity of the solvent depends mainly on the type of the solvent, the system temperature, and the amount of phenol in the wastewater [50].

The adsorption method has been found to be successful for the removal of phenols from wastewaters for a large concentration scale, depending on the adsorbent, recycling, and economics. Among the used adsorbents, activated carbon (AC) is the most preferred one in the industry. It is expensive but has been shown to be efficient for the removal of even very low amounts of organic pollutants [51–53].

Chemical oxidation processes can turn phenolic compounds into smaller molecules that are less toxic and easier to process or mineralize [54, 55]. Among the chemical oxidation processes, advanced oxidation processes (AOPs) such as the Fenton process, ozonation, photolysis, or their combinations are recommended for low-concentration wastewaters. Incineration, on the other hand, is suitable for wastewaters for which the COD values are higher than 100 g/L, however, it is no more commonly used since it is not an eco-friendly process [56].

Wet air oxidation (WAO) is a very clean technology since no additive is added to cause secondary pollution. The reaction is carried out at moderate temperatures (175–320°C) but at high pressures (2.17–20.71 MPa); the organic pollutants in the wastewater are oxidized into small organic acid molecules which are likely to biodegrade [57]. This method can be used for the treatment of wastewaters with initial COD values in the concentration range of 20–200 g/L [58].

Catalytic wet air oxidation (CWAO) offers an alternative path to treat refractory wastewaters. CWAO gained a lot of interest over the past 20 years due to its ability to oxidize toxic wastewaters and complete their mineralization [59–63]. In addition, it is a heterogeneous process, thus, an extra catalyst separation step is not necessary in most cases making the process more economical to apply [64].

Many researchers have studied the electrochemical oxidation of phenolic compounds [65, 66]. In this process, the electrode should be electrochemically stable, economically viable, and very efficient for the removal of organic pollutants [67]. There are different researches carried out using several anodic materials like Ti = SnO_2, Pt [68], vitreous carbon [69], and PbO_2 [70]. Among them, PbO_2 electrodes have been successfully applied due to their high electrical conductivity values, strong oxidizing properties, and low costs [71, 72].

Biological treatment or biodegradation is the most widely employed method for the removal of phenols from water systems. The treatment is inexpensive, simple to design and maintain, and transforms the phenolic solutions into simple end products. Phenolic molecules such as Bisphenol A (BPA) can also be successfully treated with biological treatments such as activated sludge [73–75].

CPs, on the other hand, can be removed from wastewaters by a variety of methods including biological treatment [14], advanced oxidation processes [76], and adsorption [77–79].

Although various traditional and advanced methods are possible to apply for the removal of phenol from wastewaters; the two important parameters which define the suitable method are the initial and final phenol concentrations as seen in **Table 1** [42].

Initial Concentration (mg/L)	Method	Initial Concentration (mg/L)	Method	Final Concentration (mg/L)	Method
Over 3000-6,000 or 20-500	LLE	~50	Photocatalytic Oxidation	<10	Bio-degradation
Up to 4,000 or 300-3000	Adsorption	~50	Chemical-Electrochem. Oxidation	Very low	Adsorption
~3,000	Distillation	<50	Bio-degradation	Very low	Photocatalytic Oxidation

Table 1.
Initial and Final Phenol Concentrations of Waste Waters and Corresponding Removal Methods (Modified from [42]).

6. Comparison of different methods for the removal of phenol and chlorophenols

Researchers still extensively focus on phenol removal methods from wastewaters, considering both traditional methods such as adsorption and steam distillation and advanced processes, such as wet air oxidation and biodegradation. The traditional methods mostly have the drawbacks of low efficiency and high operational cost which can be cured by using low-cost adsorbents or increasing the surface area of the existing adsorbents. In the case of advanced methods, enzymatic treatment which uses different peroxidases seems to be an efficient method with removal efficiencies above 95% [49].

As for CPs, adsorption, biodegradation, and oxidation by AOPs seem more widely used than other methods [76–79]. AOPs involve the formation and use of hydroxyl ions (OH^-) through chemical, photochemical, or photocatalytic methods [77]. Adsorption has been reported to be one of the most successful methods for CP removal from wastewaters since it is a simple method with a low-maintenance cost, high efficiency, and less toxic by-product generation [79].

7. Conclusions

Phenols and CPs are important compounds used in the manufacture of epoxy resins, plastics, polycarbonates, nylon, dyes, disinfectants herbicides, drugs, pesticides, wood preservatives, and fungicides among others. However, they are quite toxic chemicals, even at low concentrations, causing harmful effects ranging from sour mouth, diarrhea, and impaired vision to histopathological alterations, genotoxicity, mutagenicity, carcinogenicity, and death. Thus, wastewaters containing phenol and CPs have to be treated thoroughly before being discharged into water systems. The water treatment processes used in industry depend on the type of wastewater and initial and final concentrations of the phenol and CPs. Traditional methods such as adsorption and distillation or advanced methods such as wet air oxidation and biological treatment can be used to remove phenol and CPs from wastewaters. For the removal of phenol from wastewaters, among traditional methods, the adsorption method is efficient on a large scale of concentrations, depending on the economics, recycling, and the adsorbent properties. Among advanced methods, on the other hand, the inexpensive and simple to design biological treatment is the most commonly applied method and the enzymatic treatment yields more than 95% phenol removal efficiency using different peroxidases. Finally, for the removal of chlorophenols from wastewater systems, the most used methods are adsorption, biological treatment, and advanced oxidation processes.

Author details

Yesim Gucbilmez
Department of Chemical Engineering/ESTU, Eskisehir, Turkey

*Address all correspondence to: ygucbilmez@eskisehir.edu.tr

References

[1] Tang W. Physicochemical Treatment of Hazardous Wastes. 1st ed. Florida: Routledge Taylor & Francis Group, CRC Press; 2016. 608 p. ISBN: 9781566769273

[2] Taamalli A, Contreras M, Abu-Reidah I, Trabelsi N, Ben YN. Quality of phenolic compounds: Occurrence, health benefits, and applications in food industry. Journal of Food Quality. 2019;**2019**:1-2. DOI: 10.1155/2019/9594646

[3] Laura A, Moreno-Escamilla J, Rodrigo-García J, Alvarez-Parrilla E. Phenolic compounds. In: Yahia E, Carillo-Lopez A, editors. by. 1st ed. Woodhead Publishing; 2018. Ceccaroli B, Lohne O. Solar grade silicon feedstock. In: Luque A, Hegedus S, editors. Handbook of Photovoltaic Science and Engineering. 2nd ed. Chichester: Wiley; 2011. pp. 169-217. DOI: 10.1002/978047974704.ch5

[4] Anku W, Mamo M, Govender P. Phenolic compounds in water: Sources, reactivity, toxicity and treatment methods. In: Soto-Hernández M, Palma-Tenango M, García-Mateos R, editors. Phenolic Compounds. London: InTechOpen; 2017. DOI: 10.5772/67213

[5] Davì M, Gnudi F. Phenolic compounds in surface water. Water Research. 1999;**33**(14):3213-3219. DOI: 10.1016/S0043-1354(99)00027-5

[6] Olaniran A, Igbinosa E. Chlorophenols and other related derivatives of environmental concern: Properties, distribution and microbial degradation processes. Chemosphere. 2011;**83**(10):1297-1306. DOI: 10.1016/j.chemosphere.2011.04.009

[7] Runge F. Ueber einige Produkte der Steinkohlendestillation. Annalen der Physik und Chemie. 1834;**107**(5):65-78. DOI: 10.1002/andp.18341070502

[8] Runge F. Ueber einige Producte der Steinkohlendestillation. Annalen der Physik und Chemie. 1834;**108**(20-25): 328-333

[9] Reichenbach. Meinung von dem Kyanol, der Karbolsäure u. s. f. des Hrn. Runge. Annalen der Physik und Chemie. 1834;**107**(32):497-512

[10] Nguyen MT, Kryachko ES, Vanquickenborne LG. General and theoretical aspects of phenols. In: Rappoport Z et al., editors. The Chemistry of Phenols. Chichester: Wiley; 2013. p. 6. DOI: 10.1002/0470857277

[11] Reddy N, Elias A. Chlorine and the chemistry of disinfectants. Resonance. 2021;**26**(3):341-366. DOI: 10.1007/s12045-021-1135-y

[12] Kenyon R, Boehmr N. Phenol by sulfonation. Industrial and Engineering Chemistry. 1950;**42**(8):1446-1455

[13] Mukhopadhyay AK. Industrial Chemical Cresols and Downstream Derivatives. New York: Marcel Dekker; 2005. 202 p.

[14] Shinohara Y. Development in the manufacturing process of phenol. Journal of Synthetic Organic Chemistry, Japan. 1977;**35**(2):138-146. DOI: 10.5059/yukigoseikyokaishi.35.138

[15] Mitsui Petrochemical Industries. Preparation of Phenol Through Cumene. EP0371738A2. 1989

[16] Daowdat B, Hoeltzel G, Tannenbaum R. Direct Route to Phenol from Benzene [Internet]. 2017. Available from: https://repository.upenn.edu/cgi/viewcontent.cgi?article=1089&context=cbe_sdr [Accessed: 2021-12-25]

[17] Uyigue L, Kubiangha EO. Estimation of Product Yield and Kinetic

Parameters of Phenolic Resin Pre-Polymer Synthesis Using Acid and Base Catalyst. International Journal of Engineering and Modern Technology. 2018;**4**(3):25-39

[18] Yang G, Rohde B, Tesefay H, Robertson M. Biorenewable epoxy resins derived from plant-based phenolic acids. ACS Sustainable Chemistry & Engineering. 2016;**4**(12):6524-6533. DOI: 10.1021/acssuschemeng.6b01343

[19] Cygan M, Szemień M, Krompiec S. Statistical screening analysis of the chemical composition and kinetic study of phenol-formaldehyde resins synthesized in the presence of polyamines as co-catalysts. PLoS One. 2018;**13**(5):e0195069. DOI: 10.1371/journal.pone.0195069

[20] Kim W, Joshi U, Lee J. Making polycarbonates without employing phosgene: An overview on catalytic chemistry of intermediate and precursor syntheses for polycarbonate. Industrial & Engineering Chemistry Research. 2004;**43**(9):1897-1914. DOI: 10.1021/ie034004z

[21] Wang B, Wang L, Zhu J, Chen S, Sun H. Condensation of phenol and acetone on a modified macroreticular ion exchange resin catalyst. Frontiers of Chemical Science and Engineering. 2013;**7**(2):218-225. DOI: 10.1007/s11705-013-1330-9

[22] Vedamurthy T, Murugesan M. Synthesis, characterization, and evaluation of the hydrophobic, dielectric properties of phenols functionalized nylon 6 polymers by zinc acetate catalyst using Mannich reaction. Materials Chemistry and Physics. 2018;**216**:517-525. DOI: 10.1016/j.matchemphys.2018.05.070

[23] Im KM, Jeon J. Synthesis of plant phenol-derived polymeric dyes for direct or mordant-based hair dyeing. Journal of Visualized Experiments. 2016;**118**:54772. DOI: 10.3791/54772

[24] Ajani O, Akinremi O, Ajani A, Edobor-Osoh A, Anake W. Synthesis and spectroscopic study of naphtholic and phenolic azo dyes. Physical Review & Research International. 2013;**3**(1):28-41

[25] DeBono R, Laitung G. Phenolic household disinfectants — further precautions required. Burns. 1997;**23**(2):182-185. DOI: 10.1016/s0305-4179(96)00114-3

[26] Argese E, Bettiol C, Marchetto D, De Vettori S, Zambon A, Miana P, et al. Study on the toxicity of phenolic and phenoxy herbicides using the submitochondrial particle assay. Toxicology In Vitro. 2005;**19**(8):1035-1043. DOI: 10.1016/s0305-4179(96)00114-3. DOI: 0.1016/j.tiv.2005.05.004

[27] Michałowicz J, Duda W. Phenols – Sources and Toxicity. Polish Journal of Environmental Studies. 2007;**16**(3):347-362

[28] Rättö M, Ritschkoff A, Viikari L. Enzymatically polymerized phenolic compounds as wood preservatives. Holzforschung. 2004;**58**(4):440-445. DOI: 10.1515/HF.2004.067

[29] Zabka M, Pavela R. Antifungal efficacy of some natural phenolic compounds against significant pathogenic and toxinogenic filamentous fungi. Chemosphere. 2013;**93**(6):1051-1056. DOI: 10.1016/j.chemosphere.2013.05.076

[30] Badanthadka M, Mehendale H. Chlorophenols. In: Wexler P. et al., ed. by. Encyclopedia of Toxicology Volume 1. Amsterdam: Elsevier; 2014. ISBN: 9780080548005

[31] Igbinosa E, Odjadjare E, Chigor V, Igbinosa I, Emoghene A, Ekhaise F,

et al. Toxicological profile of chlorophenols and their derivatives in the environment: The public health perspective. The Scientific World Journal. 2013;**2013**:1-11. DOI: 10.1155/ 2013/460215

[32] Ullman F. Ullmann's Encyclopedia of Industrial Chemistry. 6th ed. New York: Wiley; 2000. ISBN: 9783527306732 3527306730

[33] Eikhman R, Shemyakin M, Vozhdaeva V. Anilinokrasochnaya. 1934;**4**:523

[34] Tishchenko V, Churbakov A. Journal of Applied Chemistry of the USSR. 1934;7:764

[35] Vorozhtzov NN Jr, Oshuev AG. Anilinokrasochnaya Promyshl. 1933;**3**:245

[36] Global Phenol Market Report and Forecast 2021-2026 [Internet]. Expertmarketresearch.com. 2021. [cited 7 September 2021]. Available from: https://www.expertmarketresearch.com/reports/phenol-market-report

[37] Chaudhry G, Chapalamadugu S. Biodegradation of halogenated organic compounds. Microbiological Reviews. 1991;**55**(1):59-79. DOI: 10.1128/ mr.55.1.59-79.1991

[38] Wade L. phenol | Definition, Structure, Uses, & Facts [Internet]. Encyclopedia Britannica. 2021. [cited 7 September 2021]. Available from: https://www.britannica.com/ science/phenol

[39] [Internet]. Shell.com. 2021. [cited 7 September 2021]. Available from: https://www.shell.com/business-customers/chemicals/our-products/ phenol/_jcr_content/par/textimage.stream/1447712147800/e27964363 11990d667d25cd639351276d2ea574d/ datasheet-phenol.pdf

[40] Yao L, Jiang Y, Shi J, Tomas-Barberan F, Datta N, Singanusong R, et al. Flavonoids in Food and Their Health Benefits. Plant Foods for Human Nutrition. 2004;**59**(3):113-122. DOI: 10.1007/s11130-004-0049-7

[41] Czaplicka M. Sources and transformations of chlorophenols in the natural environment. Science of the Total Environment. 2004;**322**(1-3):21-39. DOI: 10.1016/j.scitotenv.2003.09.015

[42] Mohammadi S, Kargari A, Sanaeepur H, Abbassian K, Najafi A, Mofarrah E. Phenol removal from industrial wastewaters: A short review. Desalination and Water Treatment. 2014;**53**(8):2215-2234. DOI: 10.1080/ 19443994.2014.883327

[43] Li D, Park J, Oh J. Silyl derivatization of alkylphenols, chlorophenols, and bisphenol A for simultaneous GC/MS determination. Analytical Chemistry. 2001;**73**(13): 3089-3095. DOI: 10.1021/ac001494l

[44] Santana C, Padrón M, Ferrera Z, Rodríguez J. Development of a solid-phase microextraction method with micellar desorption for the determination of chlorophenols in water samples. Journal of Chromatography A. 2007;**1140**(1-2):13-20. DOI: 10.1021/ ac001494l

[45] de Morais P, Stoichev T, Basto M, Vasconcelos M. Extraction and preconcentration techniques for chromatographic determination of chlorophenols in environmental and food samples. Talanta. 2012;**89**:1-11. DOI: 10.1016/j.talanta.2011.12.044

[46] Jin M, Yang Y. Simultaneous determination of nine trace mono- and di-chlorophenols in water by ion chromatography atmospheric pressure chemical ionization mass spectrometry. Analytica Chimica Acta. 2006;**566**(2):193-199. DOI: 10.1016/j. aca.2006.03.017

[47] Kulkarni S, Kaware D. Review on research for removal of phenol from wastewater. International Journal of Scientific and Research Publications. 2013;**3**(4):1-5

[48] Pérez Ramírez EE, de la Luz Asunción M, Saucedo Rivalcoba V, Martínez Hernández A, Velasco Santos C. In: Soto-Hernández M, Palma-Tenango M, García-Mateos R, editors. Phenolic Compounds Natural Sources, Importance and Applications. London: InTechOpen; 2021. pp. 344-372. DOI: 10.5772/67213

[49] Villegas L, Mashhadi N, Chen M, Mukherjee D, Taylor K, Biswas N. A short review of techniques for phenol removal from wastewater. Current Pollution Reports. 2016;**2**(3):157-167. DOI: 10.1007/s40726-016-0035-3

[50] Busca G, Berardinelli S, Resini C, Arrighi L. Technologies for the removal of phenol from fluid streams: A short review of recent developments. Journal of Hazardous Materials. 2008;**160**(2-3):265-288

[51] Chemat F, Boutekedjiret C. Extraction//Steam distillation. Reference Module in Chemistry, Molecular Sciences and Chemical Engineering. Amsterdam: Elsevier; 2021. ISBN: 9780124095472 012409547X

[52] Park H, Koduru J, Choo K, Lee B. Activated carbons impregnated with iron oxide nanoparticles for enhanced removal of bisphenol A and natural organic matter. Journal of Hazardous Materials. 2015;**286**:315-324. DOI: 10.1016/j.jhazmat.2014.11.012

[53] Tran V, Ngo H, Guo W, Zhang J, Liang S, Ton-That C, et al. Typical low cost biosorbents for adsorptive removal of specific organic pollutants from water. Bioresource Technology. 2015;**182**:353-363. DOI: 10.1016/j.biortech.2015.02.003

[54] Zhou L, Cao H, Descorme C, Xie Y. Phenolic compounds removal by wet air oxidation based processes. Frontiers of Environmental Science & Engineering. 2017;**12**(1):1-20. DOI: 10.1007/s11783-017-0970-2

[55] Ribeiro A, Nunes O, Pereira M, Silva A. An overview on the advanced oxidation processes applied for the treatment of water pollutants defined in the recently launched Directive 2013/39/EU. Environment International. 2015;**75**:33-51. DOI: 10.1016/j.envint.2014.10.027

[56] Andreozzi R. Advanced oxidation processes (AOP) for water purification and recovery. Catalysis Today. 1999;**53**(1):51-59. DOI: 10.1016/S0920-5861(99)00102-9

[57] Víctor-Ortega M, Ochando-Pulido J, Martínez-Férez A. Phenols removal from industrial effluents through novel polymeric resins: Kinetics and equilibrium studies. Separation and Purification Technology. 2016;**160**:136-144. DOI: 10.1016/j.seppur.2016.01.023

[58] Debellefontaine H, Chakchouk M, Foussard JN, Tissot D, Striolo P. Treatment of organic aqueous wastes: Wet air oxidation and wet peroxide oxidation(R). Environmental Pollution. 1996;**1996**(92):155-164. DOI: 10.1016/0269-7491(95)00100-x

[59] Fortuny A, Ferrer C, Bengoa C, Font J, Fabregat A. Catalytic removal of phenol from aqueous phase using oxygen or air as oxidant. Catalysis Today. 1995;**24**(1-2):79-83. DOI: 10.1016/0920-5861(95)00002-w

[60] Maugans C, Akgerman A. Catalytic wet oxidation of phenol in a trickle bed reactor over a Pt/TiO_2 catalyst. Water Research. 2003;**37**(2):319-328. DOI: 10.1016/s0043-1354(02)00289-0

[61] Suárez-Ojeda M, Fabregat A, Stüber F, Fortuny A, Carrera J, Font J.

Catalytic wet air oxidation of substituted phenols: Temperature and pressure effect on the pollutant removal, the catalyst preservation and the biodegradability enhancement. Chemical Engineering Journal. 2007;**132**(1-3):105-115. DOI: 0.1016/j.cej.2007.01.025

[62] Monteros A, Lafaye G, Cervantes A, Del Angel G, Barbier J Jr, Torres G. Catalytic wet air oxidation of phenol over metal catalyst (Ru,Pt) supported on TiO_2–CeO_2 oxides. Catalysis Today. 2015;**258**:564-569. DOI: 10.1016/j.cattod.2015.01.009

[63] Baloyi J, Ntho T, Moma J. Synthesis and application of pillared clay heterogeneous catalysts for wastewater treatment: A review. RSC Advances. 2018;**8**(10):5197-5211. DOI: 10.1039/C7RA12924F

[64] Serra-Pérez E, Ferronato C, Giroir-Fendler A, Álvarez-Torrellas S, Ovejero G, García J. Highly efficient Ru supported on carbon nanosphere nanoparticles for ciprofloxacin removal: Effects of operating parameters, degradation pathways, and kinetic study. Industrial & Engineering Chemistry Research. 2020;**59**(35):15515-15530. DOI: 10.1021/acs.iecr.0c02509

[65] Comninellis C, Pulgarin C. Electrochemical oxidation of phenol for wastewater treatment using SnO_2, anodes. Journal of Applied Electrochemistry. 1993;**23**:108-112

[66] Simond O, Schaller V, Comninellis C. Theoretical model for the anodic oxidation of organics on metal oxide electrodes. Electrochimica Acta. 1997;**42**(13-14):2009-2012. DOI: 10.1016/S0013-4686(97)85475-8

[67] Gu L, Wang B, Ma H, Kong W. Multi-phase electro-chemical catalytic oxidation of wastewater. Journal of Chemical Technology & Biotechnology. 2006;**81**(10):1697-1704. DOI: 10.1002/jctb.1594

[68] Bashir M, Isa M, Kutty S, Awang Z, Aziz H, Mohajeri S, et al. Landfill leachate treatment by electrochemical oxidation. Waste Management. 2009;**29**(9):2534-2541. DOI: 10.1016/j.wasman.2009.05.004

[69] Gomes L, Freitas R, Malpass G, Pereira E, Motheo A. Pt film electrodes prepared by the Pechini method for electrochemical decolourisation of Reactive Orange 16. Journal of Applied Electrochemistry. 2008;**39**(1):117-121. DOI: 10.1007/s10800-008-9649-5

[70] Gaudet J, Tavares A, Trasatti S, Guay D. Physicochemical Characterization of Mixed RuO_2–SnO_2 Solid Solutions. Chemistry of Materials. 2005;**17**(6):1570-1579. DOI: 10.1021/cm048129l

[71] Zhao T, Lu J, Hu C, Zhu C, Zhao J, Dong W. Electrochemical degradation of 4, 4′-(propane-2, 2-diyl) diphenol in water with CeO_2=β-PbO_2=Ti electrode. International Journal of Electrochemical Science. 2014;**9**:2354-2366

[72] Stucki S, Kötz R, Carcer B, Suter W. Electrochemicalwastewater treatment using high overvoltage anodes. Part II: Anode performance and appliations. Journal of Applied Electrochemistry. 1991;**21**:99-104

[73] Busca G, Berardinelli S, Resini C, Arrighi L. Technologies for the removal of phenol from fluid streams: A short review of recent developments. Journal of Hazardous Materials. 2008;**160**(2-3):265-288. DOI: 10.1016/j.jhazmat.2008.03.045

[74] Saratale R, Hwang K, Song J, Saratale G, Kim D. Electrochemical oxidation of phenol for wastewater treatment using Ti/PbO_2 electrode. Journal of Environmental Engineering. 2016;**142**(2):04015064. DOI: 10.1061/(ASCE)EE.1943-7870.0001007

[75] Ramírez E, Asunción M, Rivalcoba V, Hernández A, Santos C.

Removal of phenolic compounds from water by adsorption and photocatalysis. In: Soto-Hernández M, Palma-Tenango M, García-Mateos R, editors. Phenolic Compounds - Natural Sources, Importance and Applications. London: Intech Open; 2017. DOI: 10.5772/66895

[76] Pera-Titus M, García-Molina V, Baños M, Giménez J, Esplugas S. Degradation of chlorophenols by means of advanced oxidation processes: a general review. Applied Catalysis B: Environmental. 2004;**47**(4):219-256. DOI: 10.1016/j.apcatb.2003.09.010

[77] Dąbrowski A, Podkościelny P, Hubicki Z, Barczak M. Adsorption of phenolic compounds by activated carbon—a critical review. Chemosphere. 2005;**58**(8):1049-1070. DOI: 10.1016/j.chemosphere.2004.09.067

[78] Soto M, Moure A, Domínguez H, Parajó J. Recovery, concentration and purification of phenolic compounds by adsorption: A review. Journal of Food Engineering. 2011;**105**(1):1-27. DOI: 10.1016/j.jfoodeng.2011.02.010

[79] Kuśmierek K. The removal of chlorophenols from aqueous solutions using activated carbon adsorption integrated with H_2O_2 oxidation. Reaction Kinetics, Mechanisms and Catalysis. 2016;**119**(1):19-34. DOI: 10.1007/s11144-016-1039-0

Chapter 8

Occurrence and Removal of Persistent Organic Pollutants (POPs)

Siyabonga Aubrey Mhlongo, Linda Lunga Sibali, Kholofelo Clifford Malematja and Peter P. Ndibewu

Abstract

Since the revelation in the detection of the persistent organic pollutants (POPs) in industrial wastewater in the early 1990s, a notable progress has been achieved on the research and different removal applications or methods of this challenge at hand. This book chapter entails a decent understanding on the occurrence, effects, and amputation of POPs in the water sector in advancement of municipal performances of treating industrial wastewaters and environment at large. This current chapter also presents an overview of research associated to the amputation of persistent organic pollutants (POPs) from various water bodies, i.e., river sediments, sewage plants, industrial sludges, and wastewater. Also, discussing the relationships with actual pre-treatment and removal rates. Vital characteristics such as the wastewater matrix, location, sources of POPs, materials and modules, operational parameters and problems are presented with a clear focus on removal of these organic pollutant's different sources (like, textile wastewater). The particular methods to the removal of POPs can be associated with the application of ultrafiltration, nanofiltration and reverse osmosis as advanced treatment stages are considered in correlation with the textile wastewater characteristics and removal efficiencies requirements. This gives significance to the amalgamation of physico-chemical and biological treatment with membrane processes which is likely to represent an efficient solution for the removal of POPs from textile wastewater. However, since membrane fouling and hydrophilicity are apparent in the execution of this process, this chapter also covers the effective strategies like fabrication of membrane with a suitable additive to counterattack these challenges, which are often used in membrane technological research. This chapter also proposes an updated understanding of fouling and improvement of membrane properties.

Keywords: persistent organic pollutants (POPs), ultra-filtration (UF) membranes, blending, fouling, hydrophilicity

1. Introduction

There has been an advanced progress regarding the persistent organic pollutants (POPs) - i.e., elongated-lived, lethal organic composites such as PCBs, PAHs, OCPs and dioxins which have predominantly pursued their way into the environmental sector - constitute the theme of a research programme launched in the early 1990s by the Swedish Environmental Protection Agency (SEPA). This environmental programme has raised funds estimated to SEK 50 million for the research facilities in Sweden to bring focus only on persistent pollutants [1]. Equally concerning, the data obtained by World Health Organization (WHO) in late 2016, an estimated that 1.2 billion public does not have access to clean water [2, 3]. The occurrence of persistent organic pollutants (POPs) in river water and water treatment plants has raise serious concerns, especially due to the high costs and energy consumption that comes with mitigation of these challenges – because it involves variety of steps, and over thirty processes have been primarily used [4, 5]. The apparency or the occurrence of POPs in industrial wastewaters and textile industries have led to more of ecological negative effects, these includes, i.e., good taste and odor issues of the downstream water supplies, and further forming foam. This results in inhibition of the natural self-purification processes, and worse case - negative effects on the marine life and living organism in the society [6].

Persistent organic pollutants (POPs) remains nothing else but a bunch of different chemical compounds that constitutes of different pedigrees but have common traits, viz., semi-volatility, hydrophobicity, bio-accumulative, high toxicity, and alarming persistency in the environment, and they can also drift into food chains [7]. Research have indicated major contributors of POPs in the environment, these are typically chemical industry, textile industry [8], pulp and paper industry [9], and treatment of landfill leachate [10, 11].

1.1 Textile industries contribution to POPs in waste streams

During the early months of 1990, several studies reported on textile industrial sector being the major contributor of POPs, and worse, discharging very high absorptions of different hazardous POPs, i.e., polychlorinated dibenzo-p-dioxins (PCDDs) and polychlorinated dibenzofurans (PCDFs) [12–14]. A bigger portion of polychlorinated dibenzo-p-dioxins and polychlorinated dibenzofurans from textile industry are sourced all through - washing into sewage sludge, which is often used as an agricultural fertilizer, and are also the source of dioxins in the food chain [15]. A Few of these POPs like polychlorinated biphenyl, phenols, benzenes and dichloro-diphenyl-trichloroethane (DDT) are purposely formed in different of commercial applications for their significant nature or properties they have intermediates or pesticides **Table 1**.

Persistent organic pollutants are toxic chemicals which belong to the families of chemicals such as aliphatic and polycyclic aromatic hydrocarbons (PAHs), polychlorinated biphenyls (PCBs), organochlorines (OCPs), and organophosphorus pesticides (OPs) [17]. They tend to accumulate in the environment and have shown to resist photolytic, chemical, and biological degradation [18]. Persistent organic pollutants have been described by Stockholm convention as a wide range of chemicals which poses a greater risk to human life and biota due to their toxic, persistent and bio-accumulative nature [19]. Their exposure may lead to birth defects, dysfunctional immunity, change on the reproductive and/or nervous systems [20]. Therefore, their continuous detection in the environment even in low concentrations has been a genuine concern for many years. This chapter aims at giving a deep understanding on the occurrence of POPs, their nature and detection methods.

1.2 Occurrence of persistent organic pollutants (POPs)

The industrial application of POPs can be traced back during the early 1900s when these chemicals were commercialized, and used for pest and diseases control [7]. For a number of years, researchers have focused their attention on studying the

POPs	Effects	Processes or methods
Aromatic amines	Cancer-causing effect	Textile industry, dyeing
Dioxazine	Hazardous effect	Textile industry, dyeing
Antraquinone	Oncogenic effect	Textile industry, dyeing
Pentachlorophenol	Carcinogenic effect	Processing of cotton
Chloranil	Carcinogenic effect	Textile industry, dyeing
Phthalocyanine	Oncogenic effect	Textile industry, dyeing
Phenolic compounds	Harmful effects on neuraxins, liver and kidney, cancer-causing effect	Textile industry, dyeing

Table 1.
Processes and effects of some persistent organic pollutants from textile industry assembled by Mustereț and Teodosiu [16].

persistence factor, bioaccumulation, and toxicity of the common POPs such as, viz., PCBs (polychlorinated biphenyls, PAHs (polycyclic aromatic hydrocarbons) and OCPs (organochlorine pesticides) [21]. Although many POPs have been prohibited due their adverse effects however, they are still detected in considerable levels in the environment around the globe [22]. Several studies around the research space have reported significantly elevated levels of POPs in various matrices, including biota, sediment, soil, surface water, and drinking water [23–25]. It is no doubt that the rapid increase in human population, urbanization and industrialization have had a great impact in the rapid increase of the POPs in the environment [26]. Moreover, farming practices such as discharge of pesticides and fertilizers into the environment also lead to significant increase of POPs in the environment [27].

1.2.1 Compositional patterns and properties of different POPs

i. Sources of PCBs, their toxicity and nature in the environment

Polychlorinated biphenyls (PCBs) have been identified as a group of chlorinated organic pollutants consisting of 209 isomers and congeners that resulted from the variation in number and position of the chlorine atoms connected to the biphenyl rings [28, 29]. Most of these chemicals which are synthetic, have been used as coolants and lubricants mainly in electrical equipment such as electrical capacitors, generators and transformers owing to their insulating properties [30, 31].

They are characterized as persistent pollutants due to their low water solubility, high fat solubility, resistance to degradation and bioaccumulation in the environment [32]. The major concern associated with PCBs is their high level of toxicity even in extreme low concentrations. Despite their prohibition and also classified as one of the "dirty dozen" in the grouping of POPs, they are still detected in the different environment matrix [33]. Research conducted on monitoring of PCBs in the environment show that sediments are the major sources of PCBs [26]. This is because POPs such as PCBs have high organic carbon partition coefficients (Koc), making them to easily adsorb to sediments. Polycyclic biphenyls are often discharged into the environment from industrial discharge, storage leaks, volatilization, urban discharge [34].

ii. Sources of PAHs, their toxicity and nature in the environment

Polycyclic aromatic hydrocarbons are a group of lipophilic chemicals which exist in the environment in different forms (colorless, white, or yellow solids). These chemicals exist as a mixture containing two or more benzene rings fused together in linear, cluster, and angular arrangements as shown below in **Figure 1** [35]. They have been listed under Stockholm convention as POPs due to their bio-accumulative and toxic nature in the environment, while they also have been found to exhibit toxic properties such as; carcinogenic, mutagenic and teratogenic making them harmful to human health and aquatic life [36]. Naturally, PAHs can be produced from incomplete combustion of renewable materials (e.g. wild fires) [37], and volcanic eruptions [38]. However, literature shows that anthropogenic activities such as garbage burning, coal combustion, exhaust from motor vehicles, etc. dominate the sources of PAHs in the environment [39]. The persistency of PAHs tends to increase with increasing molecular weight **Figure 2** [40].

iii. Sources of OCPs, their toxicity and nature in the environment

Organochlorine pesticides are chemicals often used in agricultural activities mainly for pest control purposes [41]. They have been listed as POPs owing to their toxic, bioaccumulation and non-biodegradability nature [42]. Improper disposal from domestic use such as indoor residual spraying of pesticides plays a significant role in the increased levels of OCPs in the environment [43]. The increasing demand of agricultural practices and the persistent fight against pests mean more pesticides residues produced **Figure 3**.

According to a study by Jayaraj, Megha [44], only 0.3% of the pesticides used on crops interact with the target pest while the rest becomes excess. Therefore, these chemicals end up in different environment matrix including soil, sediments, and air. Of all the environmental matrix contamination, sediment contamination has reported to have detrimental effect on the source of food chain [45]. Furthermore, literature shows that considerable levels of OCPs have been detected in various honey samples [46–48], which is a proof of the impact that OCPS have on the food chain.

iv. Removal of persistent organic pollutants (POPs) in wastewater

Due to the continuous released of POPs into the environment, this has prompted researchers across the globe to find solutions for treating POPs. Physico-chemical methods such as coagulation, ion exchange, oxidation and adsorption have over many years been applied for removal of wide variety of POPs in the environment [18, 49]. However, many of these methods have been associated with several setbacks such as high cost. Equally important, POPs have been reported to be resistant to physico-chemical methods such as flocculation, coagulation, filtration, and oxidation process [50]. More so, bioremediation has proved to have more advantages over some physico-chemical methods due to its cost effectiveness, wide variety of the microorganisms or bio-sorbents and non-destruction of the material site [51–54]. The *in situ* bioremediation process which involves carrying out treatment

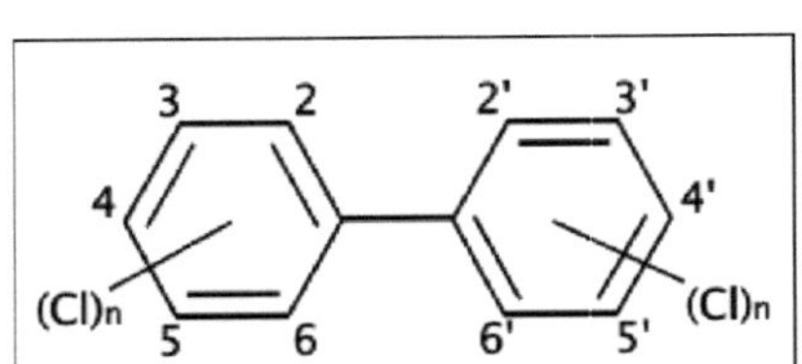

Figure 1.
Structure of PCBs.

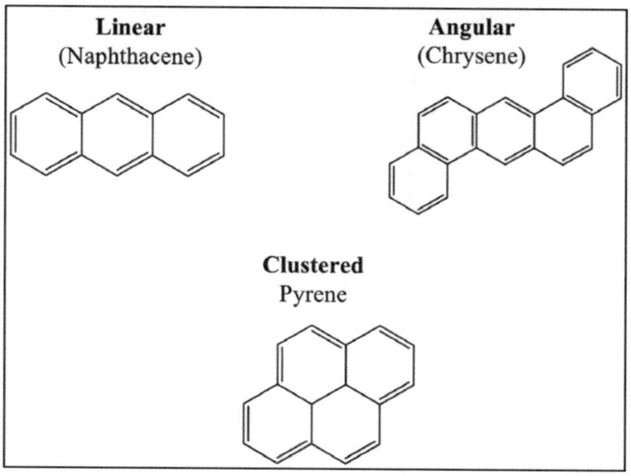

Figure 2.
Different arrangements of PAHs [35].

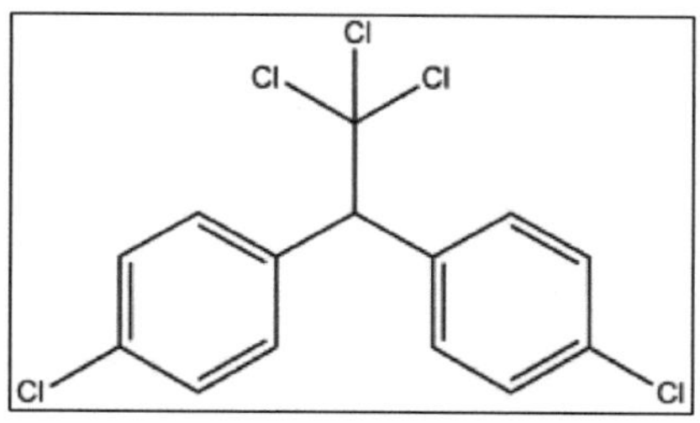

Figure 3.
Chemical structure of OCPs.

process at the contaminated site, has been regarded as a cheap, non-destructive and reliable method for degrading POPs in polluted sites [55].

Advanced oxidation process, defined as oxidation process in which hydroxyl radicals acts as oxidants, has drawn considerable recognition as a potential method for treating POPs in the environment [56]. An advanced oxidation process such as heterogeneous photocatalysis, has been widely used in degradation of POPs in the environment due to its cost effectiveness, wide availability and non-toxic properties [57]. In this heterogeneous photocatalysis, the decomposition and mineralization of contaminants using TiO_2 as photocatalyst is based on the principle of the separation of light-induced electrons/holes (e^-/h^+) pairs [58].

1.3 Removal of persistent organic pollutants (POPs) in wastewater membrane method

Membrane technology have caught so much attention in the research sector due to the drastic growth over a short space of time. This is due to its approach with advantages of using reasonable energy, less chemical matrix, good film forming ability, flexibility, robustness, separation properties, and recently, they can easily integrate with a number of methods [59, 60]. Ultrafiltration (UF) membranes are likely to be the approach having replaced macromolecular separation technique such as proteins – and apart from being the newest approach, UF membranes have some very good attributes like, low energy consumption, mild operating conditions, no phase change and they are environmentally friendly [61]. There are many polymeric materials that have been used before in the membrane processes, however, poly(ether)sulfone (PES) is mostly preferred material in (UF) membranes because of excellent properties (mechanical, thermal, and chemical stability). Some of the famously highlighted challenge about PES is the factor of hydrophobicity. This shortcoming have led to the announcements of membrane fouling from previously reported studies [62, 63]. The other polymer material previously investigated are cellulose [64], poly(vinylidene fluoride (PVDF) [65, 66], polyetherimide [67, 68], polysulfone (PS) [69] and polyethersulfone (PES) [70]. Nevertheless, PES remains the preferable membrane materials in the synthesis of UF membranes, for decades because of its convenient features.

Now, surface modification of polymeric membranes can be physical, chemical, or said to be bulky modified (i.e., polymer blends) [71]. Any type of membrane modification, be physical or chemical method - after the membrane is formed, it creates a more hydrophilic surface. These vast modification techniques can be classified into three processes, (i) graft polymerization, this is when smaller particles with hydrophilic nature are smoothly distributed or chemically infused onto the membrane scaffold; (ii)physical pre-adsorption of hydrophilic components to the membrane surface plasma treatment, this is slightly different because, there is rather a selected or a change of a functional group to the membrane surface [i.e., sulphonation, carboxylation, etc]; and (iii) Former studies confirms different kinds of modification procedures for the modification of PES membranes, namely, physical methods like blending and surface-coating methods [72, 73], and chemical methods including photo-induced grafting [74], and plasma treatment and plasma-induced grafting [75, 76].

1.3.1 Membrane blending method as an effective technique for the removal of POPs

1.3.1.1 Blending method

Usually, for an improved PES polymer property, blending method should be taken into considerations because of its simplicity and efficiency it has shown over the years. In order to observe a noticeable change in the performance of the membrane, blending method should be a necessity – this is when both PES polymer is mixed together with poly vinyl pyrrolidone (PVP), and thawed in *N*-methyl-2-pyrrolidone (NMP). The resultant polymer resin formed from the mixture should be left to be stable until handled further as normal casting technique [77]. Nearly, the idea of blending is to consortium or improve a material in a hydrophobic nature into a good mechanically hydrophilic material. This is achieved by directly blending a hydrophilic polymer like as PVP [78, 79] and poly (ethylene) glycol (PEG) [80], in that way, PES membranes are easily modified.

In this case, PVP is considered for the formation of micropores, in that way, the hydrophilicity and the antifouling properties of the membrane are increased [80]. Therefore, polymer blending technique gives rise to polymeric membrane with much improved performances and improved properties in reference to the pristine or bare PES membrane. Some researchers have encountered significant shortcomings, based on miscibility of the polymer [81]. In one way or other, there are going to be unexpected challenges with the miscibility which is limited to a narrow concentration range of vinyl pyrrolidone. These challenges are eventually resolved by blending sulfonated PES with the original PES, this is what has been done before [82, 83]. This positively outcomes the higher water permeability, and high rejections in the synthesized membrane – hence the confirmation by the sudden appearance of smaller pore sizes [84, 85]. Hence a clear indication that hydrophilicity can be wide-ranging by changes in the composition ratios of blending.

1.3.2 Considerations affecting the removal of POPs by NF/UF PES membranes

1.3.2.1 The membrane characteristics

Throughout the process of eliminating POPs from the source of waterbodies (or wastewater samples), PES material membranes become a vital factor if you consider the nature of the apparent POPs. This accurate selection of a PES membrane largely plays a role in the removal mechanism since the process is strongly related to the type and functional groups in the membrane chosen. Subsequently, there is also a significant aspect to contemplate in a suitable membrane selection, and that is - the molecular weight cut-off (MWCO), normally articulated in Dalton. This indicates the molecular weight of a hypothetical non-charged solute lying between 85 and 90% rejection, the porosity of a membrane, the surface charge, and the membrane material (polymer composition) as well as the degree of ionic species rejection [86]. In conclusion, the effect of each constraint on the removal of POPs is specifically related to the actual solute properties (molecular weight, molecular size, acid disassociation constant-pKa, and hydrophobicity/hydrophilicity — $logK_{ow}$), with which this governs the strength of the POPs-membranes physical and chemical interactions.

1.3.2.2 Membrane charge

Usually referred to as zeta potential, membrane surface charge is another vital factor to primarily study in membrane properties. The fundamental principle of the above factor lies in the fabrication of the membrane where you have to consider if the membrane has either a negatively or a positively charged surface [87]. Sometimes a membrane is pre-known to reject negatively charged pollutants (in this case, anions), such as nitrates, sulphates, and sulphites, henceforth, these fictional membranes should be negatively charged for them in order to be effectively repel the pollutants. This phenomenon therefore results into a reduced membrane fouling [88, 89]. This genius analysis of a membrane charge was discovered PVP micro particles were dispersed onto PES membrane for the membrane to give rise into an increased water permeability [90]. Thus, the zeta potential could result to many functional groups, such as, O=S=O, that comes with PES, and O=C-N of PVP that was dispersed across the scaffold of the membrane. The practical functional groups become the primary source of a negatively charged membrane [91, 92]. Consequently, this boldly confirms that an increase in the PVP particles likely to increase the hydrophilicity of the synthesized membrane – which by default leads to high permeability. Hence, the membrane charge increases as the PES and PVP dosages are varied.

1.3.2.3 Persistent organic pollutants (POPs) hydrophobicity or hydrophilicity

Hydrophilicity and hydrophobicity extremely defines the adsorption on the rejection of POPs during membrane applications process [93, 94]. Studies clearly shows that the interface between the non-polar hydrocarbon segments of POPs and the used membrane is primarily the cause of hydrophobic bonding - this has advanced the membrane progress on the extensive adsorption of POPs and of other organic pollutants onto the membrane technology [94–96]. A book published in the early 2000s vividly show that beyond hydrophobic interactions, adsorption could possibly occur over hydrogen bonding between the organic molecules and the hydrophilic groups of the membrane material [97]. Henceforth, hydrogen bonding and hydrophobic interactions may both occur independently or concurrently. Therefore, according to the studied POPs in this book, the literature approves that the hydrophobic interactions is the driving force for the organic pollutants adsorption on the membrane surfaces - this constitutes the primary step of the rejection mechanism as Nghiem and Schäfer [97] have indicated in his study. In both ways, these observations have implicitly concluded that the rejection of hydrophobic compounds should (by experimentation) be examined after the used membrane is saturated with the target compounds, otherwise, the rejection could be incorrectly mistook for adsorptions are misread [98].

1.3.3 Alternative method - Surface modification

Several studies have been done on surface modification of membranes, and it has shown decent suitability for PES material for the amputation of POPs in wastewater. This includes, self-assembling nanoparticles [99] and/or nanotubes in PES membranes. Nonetheless, this book chapter solely focuses on PES polymer as an adsorbent blended for the improvement of the membrane properties. Another method still to advance in the membrane technology is - surfactant modification. However, a little progress has been observed in literature and still requires more work to be reported on using PES membranes. In the late 2000s, Boussu, Van Baelen [100] showed an increased in the flux for the nanofiltration membrane of waterbodies comprising of surfactants. Lastly, a brief study confirms that hydrogen fluoride could be considered to advance the membrane performances. Fourteen [14] days of immersion in a hydrogen fluoride solution, an increased permeability was obtained without any loss of rejection capacities [101–103].

1.3.4 Factors to consider during the removal of POPs

1.3.4.1 Effect of the feed water composition

A modified membrane performance with real water normally consists of (i) solutions containing salts, (ii) other organic matters, (iii) pesticides, hence, POPs rejection value is likely to vary significantly depending on the feed water composition. Importantly so, pH of the water becomes a prominent influent in the POPs rejection. Below is a brief discussion of how pH is an important parameter as the driving force in rejection values.

Influence of water pH: pH in the rejection role is vitally imperative in these experiments – as it directly involved in the membrane surface and membrane charge because of the dissociation phenomenon of functional groups throughout the adsorption of POPs. Different researchers have found membrane charge (zeta potential) suddenly leaning more to negative charge whilst the pH of the water body is increased, thus, resulting in functional group deprotonation [104–106]. Moreover, another prominent researcher, Freger, Arnot [107] verified about the varying of the pore sizes likely to take place reliant on the electrostatic interactions amongst the dissociated functional groups within the membrane material. Pang, Gao [108], also showed a study where high pH ranges seemed to cause reduced rejection rates, with permeate flux also going up. And this this was ascribed by the increased number of pore sizes at high pH values. These tests were conducted during the removal of one of the top four POPs (PCBs, OCPs and DDT, etc), and expectedly, the outcome exhibited that membrane rejection achieved the highest value at pH 7, and repeatedly gave lower rejection values at pH 2.5 [109]. This clearly indicates, the ion adsorption on the membrane scaffold, and predominantly at higher pH, OH^- ion adsorption is increased - which automatically leads to an increase in the zeta potential of the membrane.

1.3.4.2 Effect of membrane fouling

Fouling of a membrane is apparent and continue to be a challenge within the membrane technology scope and industrial applications – this includes the wastewater treatment processes [86, 110]. This takes place as the undesirable particles accumulate to cause clogs in the water flow across the membrane. This results in the shortening of membrane life. Membranes looks at advancing the progress by creating membranes with better or improved properties – this means fabrication or modification to create low fouling propensity. The achievement relies on transformation of hydrophobic polymers into hydrophilic nature [111, 112]. However, it is of emergency that cost-efficiency efforts be applied in order to mitigate membrane fouling as much as possible. For this to be counter-attacked effectively, mechanisms of membrane fouling should be studied expansively, hence, to develop dynamic anti-fouling methodologies.

2. Conclusions

The contamination regulator of persistent organic pollutants (POPs) due to industrial and textile discharged effluents has become more severe and, clearly demands for interventions of more efficient wastewater advanced treatment. This leads to a combination of physico-chemical and biological treatment using membrane methods – which in-fact, embodies an efficient solution for the removal of POPs from these industrial and textile wastewaters.

In conclusion, application of membrane methods could successfully rely on several factors for its optimum use, i.e., material composition membrane selection, type of modules, wastewater characteristics and the interactions between contaminants (POPs) and the synthesized membrane. Membrane procedures potential use, for the removal of significance organic pollutants in industrial water bodies and from textile effluents are ultrafiltration (UF), reverse osmosis (RO) and nanofiltration (NF). These employed methods are cost-effective and easier to carry out. However, for fiscal and monetary reasons, these applications remain a disadvantage in the case where the effluents or wastewaters can be recuperated for re-use.

Application of PES polymeric membranes for these procedures or the removal of POPs contains increased removal rates, and the choice of a membrane material becomes paramount important considering properties of PES like, permeability, selectivity, chemical and mechanical resistance. But PES also have some integral operational challenges, such as: fouling and concentration/polarization phenomena. This further leads to an unexpected decrease of the permeate flux, and the vital aspect in the operational procedure of PES membranes and the performance of the membrane inevitably decrease. However, washing of membrane by the use of physical and chemical procedures could discreetly recover the permeate flux between the membrane processes cycles, yet fundamentally irreversible fouling could possibly emerge.

Acknowledgements

The authors would like to firstly appreciate the IntechOpen for the invitation to collaborate on this book, and also the editors, Dr. Mohamed Nageeb Rashed. Not forgetting to thank National research foundation (NRF) for funds to carry out this research (Ref No. MND190619448884). A Tshwane University of Technology (TUT) chemistry team for support and encouragement. Special thanks to supervisors, Prof Peter P. Ndibewu and Prof Linda L. Sibali for their outstanding guidance and exposing us to the greater side of science.

Conflict of interest

The authors declare no conflict of interest.

Author details

Siyabonga Aubrey Mhlongo[1]*, Linda Lunga Sibali[1], Kholofelo Clifford Malematja[2] and Peter P. Ndibewu[2]

1 Department of Environmental Sciences, University of South Africa, South Africa

2 Department of Chemistry, Tshwane University of Technology, Pretoria, South Africa

*Address all correspondence to: siyamhlongo1456@gmail.com

References

[1] Bernes C. Persistent organic pollutants. 1998.

[2] Supply WUJW, Programme SM. Progress on drinking water and sanitation: 2014 update: World Health Organization; 2014.

[3] Organization WH. Progress on drinking water, sanitation and hygiene: 2017 update and SDG baselines. 2017.

[4] Wetzel RG. Limnology: lake and river ecosystems: Gulf Professional Publishing; 2001.

[5] Drinan JE, Spellman F. Water and wastewater treatment: A guide for the nonengineering professional: Crc Press; 2012.

[6] Covaliov V, Covaliova O, Cincilei A, Mailhot G, Besse P, Delort A-M, et al. COMPLEX APPROACH TO THE PROBLEM OF PERSISTENT ORGANIC POLLUTANTS DEGRADATION IN WATER ENVIRONMENT. Environmental Engineering & Management Journal (EEMJ). 2004;3(4).

[7] Jones KC, De Voogt P. Persistent organic pollutants (POPs): state of the science. Environmental pollution. 1999;100(1-3):209-221.

[8] Métivier-Pignon H, Faur-Brasquet C, Le Cloirec P. Adsorption of dyes onto activated carbon cloths: approach of adsorption mechanisms and coupling of ACC with ultrafiltration to treat colored wastewaters. Separation and purification Technology. 2003;31(1):3-11.

[9] Pokhrel D, Viraraghavan T. Treatment of pulp and paper mill wastewater—a review. Science of the total environment. 2004;333(1-3):37-58.

[10] Kim Y, Ohsako M. A study on the appearance of persistent organic pollutants (POPs) in leachate treatment processes. Waste management & research. 2002;20(3):243-250.

[11] Wei Z, Xu T, Zhao D. Treatment of per-and polyfluoroalkyl substances in landfill leachate: status, chemistry and prospects. Environmental Science: Water Research & Technology. 2019; 5(11):1814-1835.

[12] Fernandez-Gonzalez R, Yebra-Pimentel I, Martinez-Carballo E, Simal-Gandara J. A critical review about human exposure to polychlorinated dibenzo-p-dioxins (PCDDs), polychlorinated dibenzofurans (PCDFs) and polychlorinated biphenyls (PCBs) through foods. Critical reviews in food science and nutrition. 2015;55(11): 1590-1617.

[13] Kanetoshi A, Ogawa H, Katsura E, Kaneshima H, Miura T. Formation of polychlorinated dibenzo-p-dioxins upon combustion of commercial textile products containing 2, 4, 4′-trichloro-2′-hydroxydiphenyl ether Irgasan® DP3000. Journal of Chromatography A. 1988;442:289-299.

[14] Horstmann M, McLachlan MS. Textiles as a source of polychlorinated dibenzo-p-dioxins and dibenzofurans (PCDD/F) in human skin and sewage sludge. Environmental Science and Pollution Research. 1994;1(1):15-20.

[15] Križanec B, Majcen Le Marechal A. Dioxins and dioxin-like persistent organic pollutants in textiles and chemicals in the textile sector. Croatica Chemica Acta. 2006;79(2):177-186.

[16] Mustereț C-P, Teodosiu C. REMOVAL OF PERSISTENT ORGANIC POLLUTANTS FROM TEXTILE WASTEWATER BY MEMBRANE PROCESSES. Environmental Engineering & Management Journal (EEMJ). 2007;6(3).

[17] Kanzari F, Syakti A, Asia L, Malleret L, Piram A, Mille G, et al. Distributions and sources of persistent organic pollutants (aliphatic hydrocarbons, PAHs, PCBs and pesticides) in surface sediments of an industrialized urban river (Huveaune), France. Science of the Total Environment. 2014;478:141-151.

[18] Trojanowicz M. Removal of persistent organic pollutants (POPs) from waters and wastewaters by the use of ionizing radiation. Science of The Total Environment. 2020;718:134425.

[19] Magulova K, Priceputu A. Global monitoring plan for persistent organic pollutants (POPs) under the Stockholm convention: triggering, streamlining and catalyzing global POPs monitoring. Environmental pollution. 2016;217:82-84.

[20] Ribeiro C, Ribeiro AR, Tiritan ME. Occurrence of persistent organic pollutants in sediments and biota from Portugal versus European incidence: a critical overview. Journal of Environmental Science and Health, Part B. 2016;51(3):143-153.

[21] Akoumianaki I, Potts J, Pohle I, Coull M, Adenkule I. Developing Scotland's shellfish water monitoring programme. CREW Report, CRW2017_03. 2018.

[22] Das S, Aria A, Cheng J-O, Souissi S, Hwang J-S, Ko F-C. Occurrence and distribution of anthropogenic persistent organic pollutants in coastal sediments and mud shrimps from the wetland of central Taiwan. PloS one. 2020;15(1): e0227367.

[23] Toan VD, Mai NT. Contamination of Selected Persistent Organic Pollutants (POPs) in Sediment of Some Areas in Vietnam. Sedimentation Engineering. 2018:131.

[24] Mwakalapa EB, Mmochi AJ, Müller MHB, Mdegela RH, Lyche JL, Polder A. Occurrence and levels of persistent organic pollutants (POPs) in farmed and wild marine fish from Tanzania. A pilot study. Chemosphere. 2018;191:438-449.

[25] Hossain MM, Islam KN, Rahman IM. An overview of the persistent organic pollutants in the freshwater system. Ecological Water Quality-Water Treatment and Reuse. 2012.

[26] Rodenburg LA, Ralston DK. Historical sources of polychlorinated biphenyls to the sediment of the New York/New Jersey Harbor. Chemosphere. 2017;169:450-459.

[27] Gui D, Yu R, He X, Tu Q, Chen L, Wu Y. Bioaccumulation and bio-magnification of persistent organic pollutants in Indo-Pacific humpback dolphins (*Sousa chinensis*) from the Pearl River Estuary, China. Chemosphere. 2014;114:106-113.

[28] Tanabe S, Kannan N, Subramanian A, Watanabe S, Tatsukawa R. Highly toxic coplanar PCBs: occurrence, source, persistency and toxic implications to wildlife and humans. Environmental Pollution. 1987; 47(2):147-163.

[29] Habibullah-Al-Mamun M, Ahmed MK, Islam MS, Tokumura M, Masunaga S. Occurrence, distribution and possible sources of polychlorinated biphenyls (PCBs) in the surface water from the Bay of Bengal coast of Bangladesh. Ecotoxicology and environmental safety. 2019;167:450-458.

[30] Wang Y, Wu X, Hou M, Zhao H, Chen R, Luo C, et al. Factors influencing the atmospheric concentrations of PCBs at an abandoned e-waste recycling site in South China. Science of the Total Environment. 2017;578:34-39.

[31] Wu Z, Lin T, Li A, Zhou S, He H, Guo J, et al. Sedimentary records of polychlorinated biphenyls in the East

China Marginal Seas and Great Lakes: Significance of recent rise of emissions in China and environmental implications. Environmental Pollution. 2019;254: 112972.

[32] Katsoyiannis A. Occurrence of polychlorinated biphenyls (PCBs) in the Soulou stream in the power generation area of Eordea, northwestern Greece. Chemosphere. 2006;65(9):1551-1561.

[33] Malina N, Mazlova EA. Temporal and spatial variation of polychlorinated biphenyls (PCBs) contamination in environmental compartments of highly polluted area in Central Russia. Chemosphere. 2017;185:227-236.

[34] Alegria H, Martinez-Colon M, Birgul A, Brooks G, Hanson L, Kurt-Karakus P. Historical sediment record and levels of PCBs in sediments and mangroves of Jobos Bay, Puerto Rico. Science of the Total Environment. 2016;573:1003-1009.

[35] Abdel-Shafy HI, Mansour MS. A review on polycyclic aromatic hydrocarbons: source, environmental impact, effect on human health and remediation. Egyptian journal of petroleum. 2016;25(1):107-123.

[36] Adekunle AS, Oyekunle JAO, Ola IJ, Obisesan OR, Maxakato NW. Determination of polycyclic aromatic hydrocarbons (PAHs) and organochlorine pesticides (OCPs) in some personal care products in Nigeria. Toxicology reports. 2018;5:994-1001.

[37] Faboya OL, Sojinu SO, Oguntuase BJ, Sonibare OO. Impact of forest fires on polycyclic aromatic hydrocarbon concentrations and stable carbon isotope compositions in burnt soils from tropical forest, Nigeria. Scientific African. 2020;8:e00331.

[38] Kozak K, Ruman M, Kosek K, Karasiński G, Stachnik Ł, Polkowska Ż. Impact of volcanic eruptions on the occurrence of PAHs compounds in the aquatic ecosystem of the southern part of West Spitsbergen (Hornsund Fjord, Svalbard). Water. 2017;9(1):42.

[39] Patel AB, Shaikh S, Jain KR, Desai C, Madamwar D. Polycyclic aromatic hydrocarbons: sources, toxicity and remediation approaches. Frontiers in Microbiology. 2020;11:2675.

[40] Cerniglia CE. Biodegradation of polycyclic aromatic hydrocarbons. Current opinion in biotechnology. 1993;4(3):331-338.

[41] Corsolini S, Covaci A, Ademollo N, Focardi S, Schepens P. Occurrence of organochlorine pesticides (OCPs) and their enantiomeric signatures, and concentrations of polybrominated diphenyl ethers (PBDEs) in the Adélie penguin food web, Antarctica. Environmental pollution. 2006; 140(2):371-382.

[42] Zhang Q, Chen Z, Li Y, Wang P, Zhu C, Gao G, et al. Occurrence of organochlorine pesticides in the environmental matrices from King George Island, west Antarctica. Environmental Pollution. 2015;206: 142-149.

[43] Olisah C, Okoh OO, Okoh AI. Occurrence of organochlorine pesticide residues in biological and environmental matrices in Africa: A two-decade review. Heliyon. 2020;6(3):e03518.

[44] Jayaraj R, Megha P, Sreedev P. Organochlorine pesticides, their toxic effects on living organisms and their fate in the environment. Interdisciplinary toxicology. 2016;9(3-4):90.

[45] López R, Goñi F, Etxandia A, Millán E. Determination of organochlorine pesticides and polychlorinated biphenyls in human serum using headspace solid-phase microextraction and gas chromatography-electron capture

detection. Journal of Chromatography B. 2007;846(1-2):298-305.

[46] Muiño MF, Sancho M, Gándara JS, Vidal JC, Huidobro J, Lozano JS. Organochlorine pesticide residues in Galician (NW Spain) honeys. Apidologie. 1995;26(1):33-38.

[47] Mukherjee I. Determination of pesticide residues in honey samples. Bulletin of environmental contamination and toxicology. 2009;83(6):818-821.

[48] Mousavi M-M, Nemati M, Nabili AAA, Arefhosseini S. Application of dispersive liquid–liquid microextraction followed by gas chromatography/mass spectrometry as effective tool for trace analysis of organochlorine pesticide residues in honey samples. Journal of the Iranian Chemical Society. 2016;13(12):2211-2218.

[49] Rashed MN. Adsorption technique for the removal of organic pollutants from water and wastewater. Organic pollutants-monitoring, risk and treatment. 2013;7:167-194.

[50] Badmus KO, Tijani JO, Massima E, Petrik L. Treatment of persistent organic pollutants in wastewater using hydrodynamic cavitation in synergy with advanced oxidation process. Environmental Science and Pollution Research. 2018;25(8):7299-7314.

[51] Ang EL, Zhao H, Obbard JP. Recent advances in the bioremediation of persistent organic pollutants via biomolecular engineering. Enzyme and microbial technology. 2005;37(5):487-496.

[52] Ding J, Chen B, Zhu L. Biosorption and biodegradation of polycyclic aromatic hydrocarbons by Phanerochaete chrysosporium in aqueous solution. Chinese Science Bulletin. 2013;58(6):613-621.

[53] Pariatamby A, Kee YL. Persistent organic pollutants management and remediation. Procedia Environmental Sciences. 2016;31:842-848.

[54] Nguyen V-H, Smith SM, Wantala K, Kajitvichyanukul P. Photocatalytic remediation of persistent organic pollutants (POPs): a review. Arabian Journal of Chemistry. 2020;13(11):8309-8337.

[55] Boudh S, Singh JS, Chaturvedi P. Microbial resources mediated bioremediation of persistent organic pollutants. New and Future Developments in Microbial Biotechnology and Bioengineering: Elsevier; 2019. p. 283-294.

[56] Li F. Study on sono-photocatalytic degradation of POPs: a case study hydrating polyacrylamide in wastewater. Organic pollutants Intech Publisher Inc., Rijeka. 2012:327-44.

[57] Shaban YA, El Sayed MA, El Maradny AA, Al Farawati RK, Al Zobidi MI, Khan SU. Photocatalytic removal of polychlorinated biphenyls (PCBs) using carbon-modified titanium oxide nanoparticles. Applied Surface Science. 2016;365:108-113.

[58] Pi Y, Li X, Xia Q, Wu J, Li Y, Xiao J, et al. Adsorptive and photocatalytic removal of Persistent Organic Pollutants (POPs) in water by metal–organic frameworks (MOFs). Chemical Engineering Journal. 2018;337:351-371.

[59] Zheng X, Zhang Z, Yu D, Chen X, Cheng R, Min S, et al. Overview of membrane technology applications for industrial wastewater treatment in China to increase water supply. Resources, Conservation and Recycling. 2015;105:1-10.

[60] Baker RW. Membrane technology and applications: John Wiley & Sons; 2012.

[61] Ulbricht M. Advanced functional polymer membranes. Polymer. 2006;47(7):2217-2262.

[62] Van der Bruggen B, Braeken L, Vandecasteele C. Flux decline in nanofiltration due to adsorption of organic compounds. Separation and Purification Technology. 2002;29(1):23-31.

[63] Noble RD, Stern SA. Membrane separations technology: principles and applications: Elsevier; 1995.

[64] Lv C, Su Y, Wang Y, Ma X, Sun Q, Jiang Z. Enhanced permeation performance of cellulose acetate ultrafiltration membrane by incorporation of Pluronic F127. Journal of membrane science. 2007;294(1-2):68-74.

[65] Srivastava HP, Arthanareeswaran G, Anantharaman N, Starov VM. Performance of modified poly (vinylidene fluoride) membrane for textile wastewater ultrafiltration. Desalination. 2011;282:87-94.

[66] Mntambo S, Mdluli P, Mahlambi M, Onwubu S, Nxumalo N. Synthesis and characterization of ultrafiltration membranes functionalised with C18 as a modifier for adsorption capabilities of polyaromatic hydrocarbons. Water SA. 2019;45(1):131-140.

[67] Tao C-T, Young T-H. Polyetherimide membrane formation by the cononsolvent system and its biocompatibility of MG63 cell line. Journal of Membrane Science. 2006;269(1-2):66-74.

[68] Hebbar RS, Isloor AM, Prabhu B, Asiri AM, Ismail A. Removal of metal ions and humic acids through polyetherimide membrane with grafted bentonite clay. Scientific reports. 2018;8(1):1-16.

[69] Wu H, Tang B, Wu P. Development of novel SiO2–GO nanohybrid/ polysulfone membrane with enhanced performance. Journal of Membrane Science. 2014;451:94-102.

[70] Mozafari M, Seyedpour SF, Salestan SK, Rahimpour A, Shamsabadi AA, Firouzjaei MD, et al. Facile Cu-BTC surface modification of thin chitosan film coated polyethersulfone membranes with improved antifouling properties for sustainable removal of manganese. Journal of Membrane Science. 2019;588:117200.

[71] Khayet M, Matsuura T. Application of surface modifying macromolecules for the preparation of membranes for membrane distillation. Desalination. 2003;158(1-3):51-56.

[72] Shi Q, Su Y, Zhao W, Li C, Hu Y, Jiang Z, et al. Zwitterionic polyethersulfone ultrafiltration membrane with superior antifouling property. Journal of Membrane Science. 2008;319(1-2):271-278.

[73] Yang Y-F, Li Y, Li Q-L, Wan L-S, Xu Z-K. Surface hydrophilization of microporous polypropylene membrane by grafting zwitterionic polymer for anti-biofouling. Journal of Membrane Science. 2010;362(1-2):255-264.

[74] Chen K-S, Tsai J-C, Chou C-W, Yang M-R, Yang J-M. Effects of additives on the photo-induced grafting polymerization of N-isopropylacrylamide gel onto PET film and PP nonwoven fabric surface. Materials Science and Engineering: C. 2002;20(1-2):203-208.

[75] Hu J, Shao D, Chen C, Sheng G, Li J, Wang X, et al. Plasma-induced grafting of cyclodextrin onto multiwall carbon nanotube/iron oxides for adsorbent application. The Journal of Physical Chemistry B. 2010;114(20):6779-6785.

[76] Li J, Chen C, Zhao Y, Hu J, Shao D, Wang X. Synthesis of water-dispersible Fe3O4@ β-cyclodextrin by plasma-induced grafting technique for pollutant treatment. Chemical engineering journal. 2013;229:296-303.

[77] Xu Z-L, Qusay FA. Polyethersulfone (PES) hollow fiber ultrafiltration membranes prepared by PES/

non-solvent/NMP solution. Journal of Membrane Science. 2004;233(1-2): 101-111.

[78] Su B-h, Fu P, Li Q, Tao Y, Li Z, Zao H-s, et al. Evaluation of polyethersulfone highflux hemodialysis membrane in vitro and in vivo. Journal of Materials Science: Materials in Medicine. 2008;19(2):745-751.

[79] Wang H, Yu T, Zhao C, Du Q. Improvement of hydrophilicity and blood compatibility on polyethersulfone membrane by adding polyvinyl-pyrrolidone. Fibers and polymers. 2009;10(1):1-5.

[80] Wang Y-q, Wang T, Su Y-l, Peng F-b, Wu H, Jiang Z-y. Protein-adsorption-resistance and permeation property of polyethersulfone and soybean phosphatidylcholine blend ultrafiltration membranes. Journal of membrane science. 2006;270(1-2):108-114.

[81] Kim JH, Kim CK. Ultrafiltration membranes prepared from blends of polyethersulfone and poly (1-vinylpyrrolidone-co-styrene) copolymers. Journal of membrane science. 2005;262(1-2):60-68.

[82] Blanco J-F, Sublet J, Nguyen QT, Schaetzel P. Formation and morphology studies of different polysulfones-based membranes made by wet phase inversion process. Journal of membrane science. 2006;283(1-2):27-37.

[83] Rahimpour A, Madaeni SS, Ghorbani S, Shockravi A, Mansourpanah Y. The influence of sulfonated polyethersulfone (SPES) on surface nano-morphology and performance of polyethersulfone (PES) membrane. Applied surface science. 2010;256(6):1825-1831.

[84] Haitao W, Liu Y, Xuehui Z, Tian Y, Qiyun D. Improvement of hydrophilicity and blood compatibility on polyethersulfone membrane by blending sulfonated polyethersulfone. Chinese Journal of Chemical Engineering. 2009;17(2):324-329.

[85] Li S, Cui Z, Zhang L, He B, Li J. The effect of sulfonated polysulfone on the compatibility and structure of polyethersulfone-based blend membranes. Journal of Membrane Science. 2016;513:1-11.

[86] Mulder M, Mulder J. Basic principles of membrane technology: Springer Science & Business Media; 1996.

[87] Al-Amoudi A, Williams P, Mandale S, Lovitt RW. Cleaning results of new and fouled nanofiltration membrane characterized by zeta potential and permeability. Separation and purification technology. 2007;54(2):234-240.

[88] Abdel-Karim A, El-Naggar ME, Radwan E, Mohamed IM, Azaam M, Kenawy E-R. High-performance mixed-matrix membranes enabled by organically/inorganic modified montmorillonite for the treatment of hazardous textile wastewater. Chemical Engineering Journal. 2021;405:126964.

[89] Kotp YH. Removal of organic pollutants using polysulfone ultrafiltration membrane containing polystyrene silicomolybdate nanoparticles: Case study: Borg El Arab area. Journal of Water Process Engineering. 2019;30:100553.

[90] Al Malek S, Seman MA, Johnson D, Hilal N. Formation and characterization of polyethersulfone membranes using different concentrations of polyvinyl-pyrrolidone. Desalination. 2012;288: 31-39.

[91] Salgin S, Salgin U, Soyer N. Streaming potential measurements of polyethersulfone ultrafiltration membranes to determine salt effects on membrane zeta potential. Int J Electrochem Sci. 2013;8(3):4073-4084.

[92] Sotto A, Boromand A, Zhang R, Luis P, Arsuaga JM, Kim J, et al. Effect of nanoparticle aggregation at low concentrations of TiO2 on the hydrophilicity, morphology, and fouling resistance of PES–TiO2 membranes. Journal of colloid and interface science. 2011;363(2):540-550.

[93] Bolzonella D, Fatone F, di Fabio S, Cecchi F. Application of membrane bioreactor technology for wastewater treatment and reuse in the Mediterranean region: Focusing on removal efficiency of non-conventional pollutants. Journal of environmental management. 2010;91(12):2424-2431.

[94] Wang T, Zhao C, Li P, Li Y, Wang J. Fabrication of novel poly (m-phenylene isophthalamide) hollow fiber nanofiltration membrane for effective removal of trace amount perfluorooctane sulfonate from water. Journal of membrane science. 2015;477:74-85.

[95] Kiso Y, Nishimura Y, Kitao T, Nishimura K. Rejection properties of non-phenylic pesticides with nanofiltration membranes. Journal of Membrane Science. 2000;171(2):229-237.

[96] Kimura K, Amy G, Drewes JE, Heberer T, Kim T-U, Watanabe Y. Rejection of organic micropollutants (disinfection by-products, endocrine disrupting compounds, and pharmaceutically active compounds) by NF/RO membranes. Journal of membrane science. 2003;227(1-2): 113-121.

[97] Nghiem LD, Schäfer A. Trace contaminant removal with nanofiltration. Elsevier; 2004.

[98] Kimura K, Amy G, Drewes J, Watanabe Y. Adsorption of hydrophobic compounds onto NF/RO membranes: an artifact leading to overestimation of rejection. Journal of Membrane Science. 2003;221(1-2):89-101.

[99] Luo M-L, Zhao J-Q, Tang W, Pu C-S. Hydrophilic modification of poly (ether sulfone) ultrafiltration membrane surface by self-assembly of TiO2 nanoparticles. Applied Surface Science. 2005;249(1-4):76-84.

[100] Boussu K, Van Baelen G, Colen W, Eelen D, Vanassche S, Vandecasteele C, et al. Technical and economical evaluation of water recycling in the carwash industry with membrane processes. Water Science and Technology. 2008;57(7):1131-1135.

[101] Muñoz MG, Navarro R, Saucedo I, Avila M, Prádanos P, Palacio L, et al. Hydrofluoric acid treatment for improved performance of a nanofiltration membrane. Desalination. 2006;191(1-3):273-278.

[102] Navarro R, González M, Saucedo I, Avila M, Prádanos P, Martínez F, et al. Effect of an acidic treatment on the chemical and charge properties of a nanofiltration membrane. Journal of Membrane Science. 2008;307(1):136-148.

[103] González M, Saucedo I, Navarro R, Prádanos P, Palacio L, Martínez F, et al. Effect of phosphoric and hydrofluoric acid on the structure and permeation of a nanofiltration membrane. Journal of membrane science. 2006;281(1-2): 177-185.

[104] Afonso MD. Surface charge on loose nanofiltration membranes. Desalination. 2006;191(1-3):262-272.

[105] Lalia BS, Kochkodan V, Hashaikeh R, Hilal N. A review on membrane fabrication: Structure, properties and performance relationship. Desalination. 2013;326:77-95.

[106] Zargar M, Hartanto Y, Jin B, Dai S. Understanding functionalized silica nanoparticles incorporation in thin film composite membranes: Interactions and desalination performance. Journal of Membrane Science. 2017;521:53-64.

[107] Freger V, Arnot T, Howell J. Separation of concentrated organic/ inorganic salt mixtures by nanofiltration. Journal of Membrane Science. 2000; 178(1-2):185-193.

[108] Pang W, Gao N, Xia S. Removal of DDT in drinking water using nanofiltration process. Desalination. 2010;250(2):553-556.

[109] Mukherjee D, Bhattacharya P, Jana A, Bhattacharya S, Sarkar S, Ghosh S, et al. Synthesis of ceramic ultrafiltration membrane and application in membrane bioreactor process for pesticide remediation from wastewater. Process Safety and Environmental Protection. 2018;116:22-33.

[110] Guo W, Ngo H-H, Li J. A mini-review on membrane fouling. Bioresource technology. 2012;122:27-34.

[111] Meng J-Q, Yuan T, Kurth CJ, Shi Q, Zhang Y-F. Synthesis of antifouling nanoporous membranes having tunable nanopores via click chemistry. Journal of membrane science. 2012;401:109-117.

[112] Li X, Sotto A, Li J, Van der Bruggen B. Progress and perspectives for synthesis of sustainable antifouling composite membranes containing in situ generated nanoparticles. Journal of Membrane Science. 2017;524:502-528.

Chapter 9

Recent Developments in the Application of Advanced Oxidative Processes for Remediation of Persistent Organic Pollutants from Water

Ifeoluwa Oluwafunmilayo Daramola and Matthew Ayorinde Adebayo

Abstract

Environmental pollution as a result of industrialization is a continuous menace. In our precious environment, Persistent organic pollutants (POPs) are constantly present and these pollutants are of great concern because of their high level of toxicity, persistency and bioaccumulation. Therefore, this chapter discusses different types and sources of POPs in the environment. The chapter also introduces Advanced oxidative processes (AOPs) and the classes of AOPs. Removal of selected POPs from aqueous solutions by AOPs, such as sulfate radical, ionizing radiation, heterogeneous photocatalysis, electrohydraulic discharge system, ozonation, and Fenton processes, were discussed. The major aim of the chapter is to make available to environmental scientists the recent developments in the removal of POPs by AOPs.

Keywords: advanced oxidation processes, persistent organic pollutants, degradation, removal, environment

1. Introduction

Today, global industrialization has resulted in the development of a variety of chemicals that, while useful, have attracted scientific attention because of their hazardous effects on humans and environment. Among these chemicals are Persistent organic pollutants (POPs) that are of serious concern because of their level of toxicity, long-persistent nature and bio-accumulation. The earth's ecology is currently being continuously contaminated by various pollutants. Pollutants of various forms are found in many locations. Some of these POPs are resistant to environmental deterioration (chemical, biological, and photolytic reactions) and exist for a lengthy period of time in our environment [1]. Persistent organic pollutants belong to a category of organic chemicals that are persistent, toxic, bioaccumulative, and are likely to have negative impacts on human health and the environment (persistent, bioaccumulative, and toxic substances) [2].

IntechOpen

Persistent organic pollutants are defined by the Stockholm Convention as carbon-based chemicals that persist in the environment for a long period and are extensively disseminated. Persistent organic pollutants originate from man-made sources associated with the production, use, and disposal of some organic chemicals. Due to their persistence, ability to bioaccumulate in tissues, long-range transportability, and severe toxicity (even at low concentrations), POPs are a serious global hazard [3]. Persistent organic pollutants can also be produced unintentionally as by-products of combustion or chemical processing. Persistent organic pollutants are released into the environment on a regular basis, whether purposefully or unintentionally. The hydrophobicity of POPs is usually linked to halogenated compounds and these pollutants have low solubilities in water and high lipophilicities. They partition aggressively to solids, particularly organic matter, in aquatic systems and soils, avoiding the aqueous phase. These chemicals partition into lipids in organisms and are stored in fatty tissue instead of entering the aqueous milieu of cells. These chemicals exist persistently in plants and animals as a result of low metabolism [4]. Some of the POPs, such as polycyclic aromatic hydrocarbons (PAHs), can be produced from natural sources, however, POPs originate from the industries that are manufacturing a wide range of goods, such as agrochemicals, solvents, and flame-retardants [5].

2. Sources and fate of POPs in the environment

There are a number of POP chemicals, coming from certain series or 'families' of chemicals. Among the important classes of POP chemicals are many families of chlorinated (and brominated) aromatics, including polychlorinated dibenzo-p-dioxins and-furans (PCDD/Fs), polybrominated diphenyl ethers (PBDEs), polychlorinated biphenyls (PCBs), and different organochlorine pesticides (for instance DDT and its metabolites, chlordane, toxaphene, among others). Some are accidental by-products of combustion or the industrial synthesis of other chemicals (e.g., the PCDD/Fs) not produced deliberately. Many POPs have been synthesized for industrial uses (e.g., PCBs, PBDEs, and chlorinated paraffins) or as agrochemicals (e.g., chlordane, Dichlorodiphenyl trichloroethane (DDT), and Lindane). Examples of more polar POPs are phenols (e.g., polyethoxylated alkylphenols which are non-ionic surfactants), and chlorinated phenols [6].

The three main causes of rising POP levels in ecosystems are industrial and agricultural activities as well as municipal populations. Sources of POPs are mainly from anthropogenic activities and can be introduced into the environment through many pathways. These pollutants can reach the environment through urban runoff, agricultural runoff, drainage system, industrial effluent, landfill leachate and deposition from atmosphere. Waste incineration, consumer goods production, transportation, energy generation, mineral and metal mining (ferrous and non-ferrous). Chemical synthesis also emits alarmingly large amounts of POPs into our precious environment [7].

It has been shown in literature that POPs can assimilate in the environment within weeks, but it will take years or decades for POPs to naturally decompose [8]. Persistent organic pollutants are known for their semi-volatility, which is a trait of their physicochemical properties that allows them to exist in the vapor phase or adsorbed on air particles, allowing for long-range movement in the atmosphere. Persistent organic pollutants are everywhere; they possess the ability to move through air, soil and water before being naturally decomposed [9]. They have been found in both industrialized and non-industrialized places, in urban and rural settings, in heavily populated and poorly populated areas, without significant human aids. Persistent organic pollutants have been measured on every continent at locations that reflect every major climatic zone and geographic sector. These include places such as the Arctic, the open oceans, deserts, and the Antarctic, where there are no substantial local sources and long-distance movement from other areas of the world is the only conceivable explanation for their presence. Polychlorinated biphenyls have been found in the air in rates of up to 15 ng/m^3 in all parts of the world; in industrialized areas, concentrations can be several orders of magnitude higher. Rain and snow have also been found to contain PCBs [10].

Global cycling of POPs under the influence of climate change primarily demonstrates that global warming promotes secondary emission of POPs; for example, temperature rise will cause POPs to be re-released from soils and oceans while melting glaciers and permafrost will cause POPs to be re-released into freshwater ecosystems. Extreme weather events around the world, such as droughts and floods, cause POPs to be redistributed due to strong soil erosion. The global transport of POPs has been considerably influenced by changes in atmospheric circulation and ocean currents. Climate change has affected marine biological productivity, affecting the ocean's POP storage capacity. The patterns of aquatic and terrestrial food chains have changed dramatically, potentially amplifying POP toxicity in ecosystems. Generally, global warming speeds up the process of POP volatilization and increases the number of POPs in the environment, while also facilitating their breakdown. The future of environmental behaviors of POPs has been forecasted using models such as G-CIEMS (Grid-Catchment Integrated Environmental Modeling System), Berkeley-Trent Global Model (BETR-Global), and Globo-POP. Governments make use of these models to analyze the influence of global warming on the fate of POPs in the environmental and, as a result, properly control POPs [11, 12].

All human beings are being exposed to POPs at some point in their life, regardless of age, tribe or location. As proved by current epidemiological evidence, it has been suggested that early-life exposure to POPs can adversely affect the development of immune and respiratory systems [13]. Persistent organic pollutants infiltrate the human system as early as infancy, and these pollutants have been detected in variable amounts in baby meals, which are popular around the world for giving nutrients to infants [14]. According to studies, the proportion of POPs in the human body increases with age, with elderly population often having the greatest amount of POPs in the system, which is due to the fact that the metabolism of elderly people is normally slow [15]. Among the diseases associated with POPs are endocrine disturbance, obesity, diabetes, cardiovascular problems, cancer, reproductive and other health-related issues [1].

The toxicity and persistence of POPs to humans have created a need to develop effective POPs' cleanup methods. In the first instance, the production and use of POPs should be controlled. To address this, the United States joined forces with the European Community and 90 other countries to sign a groundbreaking United Nations treaty in Stockholm, Sweden on the 23 May 2001, and a convention on POPs known as the Stockholm Convention on POPs was signed. The Convention entered into force on the 17 May 2004 [16]. The major aim of the Convention is to protect the environment as well as human health from POPs by controlling the usage of POPs with the view to phasing them out. The Convention requires that each party should prohibit and/or take any administrative or legal action required for the reduction/elimination of POPs production and usage, export and import, as well as to take actions to prevent or minimize POPs' release into the environment. The Convention identified twelve POPs chemicals known as the "*dirty dozen*" for intervention, and new chemicals are considered for listing at each Conference of the parties. These chemicals are presented in **Table 1**. Nine of the initial POPs are pesticides, one is an industrial chemical and two are unintentionally produced

Chemical	Intentional production and use - Pesticide	Intentional production and use – industrial chemical	Unintentional production	Management Measure
Aldrin	X			Limited usage as local ectoparasiticide; and elimination for production
Endrin	X			Elimination
Heptachlor	X			Limited usage; and elimination for production
Mirex	X			Limited usage; and exemption for production in countries that registered for exemptions
Toxaphene	X			Elimination
DDT	X			Exemption for use in the production of dicofol; and restricted usage for malaria vector control
Chlordane	X			Limited usage; and exemption for production in countries that registered for exemptions.
Dieldrin	X			Reduction in the usage for agricultural activities; and elimination for production
Hexachlorobenzene	X	X	X	Limited usage; implementation of the measures to reduce or eliminate its release from unintentional production; and exemption for production in countries that registered for exemptions
Polychlorinated biphenyls		X	X	Elimination for usage by 2025; and elimination for production
Dioxins (polychlorinated dibenzo-p-dioxins)			X	Implementation of the measures to reduce or eliminate its release from unintentional production
Furans (polychlorinated dibenzofurans)			X	Implementation of the measures to reduce or eliminate the release from unintentional production

Table 1.
The twelve convention-identified POPs: The dirty dozen.

Chemical	Intentional production and use - pesticide	Intentional Production and use – industrial chemical	Unintentional production	Management measure
Chlordecone	X			Elimination
Lindane	X			Exemption for production in countries that registered for exemptions for usage as a human health pharmaceutical
Alpha hexachlorocyclohexane	X		X	Managing of unintentionally production; and elimination
Beta hexachlorocyclohexane	X		X	Managing of unintentionally production; and elimination
Pentachlorobenzene (PeCB)	X	X	X	Managing of unintentionally production; and elimination
Tetrabromodiphenyl ether and pentabromodiphenyl ether		X		Exemption for use as articles containing these chemicals for recycling; and elimination for production
Perfluorooctane sulfonic acid (PFOS), its salts and perfluorooctanesulfonyl fluoride (PFOS-F)		X		Phasing out with acceptable purpose and specific exemptions
Hexabromobiphenyl		X		Complete elimination
Hexabromodiphenyl ether and Heptabromodiphenyl ether		X		Exemption for use as articles containing these chemicals for recycling; and elimination for production

Table 2.
The POPs listed in May 2004.

through certain industrial processes. Nine new POPs (**Table 2**) were listed in May, 2004 which includes unintentionally produced and released POPs, which result from some industrial processes [17].

Another approach to prevent the proliferation of POPs is the development of novel advanced technologies for remediation of water pollution. Water for human and animal consumption must be adequately treated to ensure excellent health, thus pollutants must be removed. Freshwater resources are exposed to a number of organic pollutants, including dyes, medicines, industrial chemicals, pesticides, and personal care products, which are discharged directly into natural water systems on a daily basis. Treatment of industrial effluents before being released into natural water bodies is crucial for effective protection of natural water resources. Some of the successful water treatment procedures used in the service and provision of industrial or municipal potable water include flocculation, coagulation, filtration, and chlorination. Conventional treatment, however, is ineffective against removal of POPs. Despite their established lipophilicity, POPs are unlikely to adsorb on organic matter, and application of treatment chemicals frequently produce undesirable intermediates, making conventional treatment methods unfruitful [18].

However, various advanced wastewater treatment technologies such as activated carbon adsorption, membrane bioreactor (MBR) and advanced oxidation processes (AOP) have been applied in the treatment of POPs to counter the difficulty in conventional methods [19].

3. Advanced oxidation processes (AOP)

Advanced oxidative processes are aqueous phase oxidation systems that produce highly efficient oxidizing agents, such as hydroxyl and sulfate radicals that are mainly generated as the dominating species, which have high ability to destroy POPs. The radicals ensure that soluble organic pollutants are effectively degraded into biodegradable and simple molecules. The solution pH, water turbidity, duration of reaction, the amount/volume of the organic component sensitive to degradation, and the presence of OH radical scavengers or activator chemicals can all have impacts on the OH radical's degradation activity. The distinctiveness of these processes is their diversified creation of extremely reactive OH radicals, which oxidize organic contaminants in wastewater non-specifically and quickly. Depending on the treatment goals and the features of the wastewater stream, AOPs can be used as a single procedure, in combination with other AOPs, or in combination with conventional treatment techniques [20]. The benefits, which include ease of operation, high performance, low selectivity, strong reproducibility, minimal by-products' generation, and total degradation of pollutants, make AOPs viable techniques for removal of POPs from aqueous effluents [21]. Classifications of AOPs for POPs removal are depicted in **Figure 1**.

Sulfate radical (SR, SO_4^-) based AOPs have recently shown promising potentials in the degradation of non-biodegradable chemicals, with peroxydisulfate

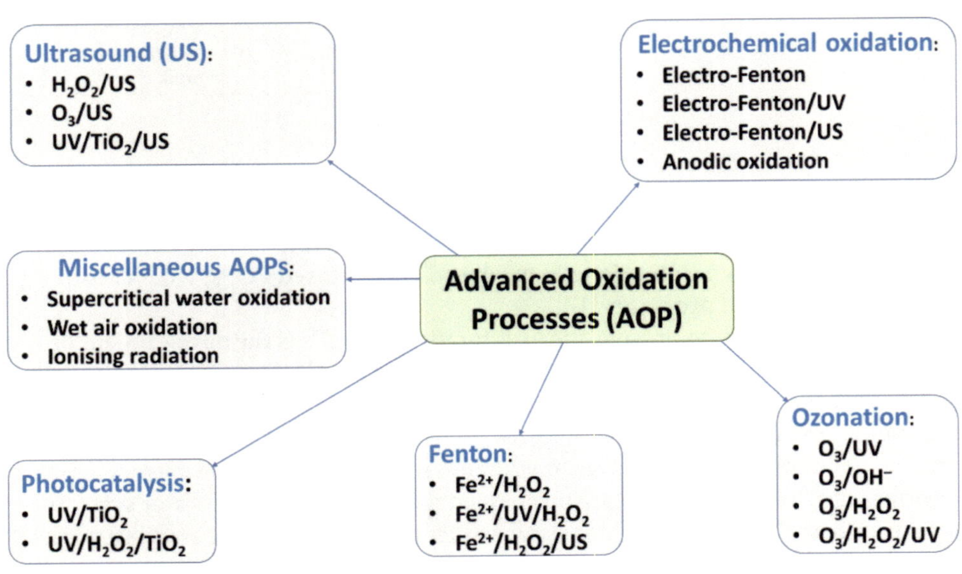

Figure 1.
Types of AOPs for POPs removal and degradation.

(PDS, $S_2O_8^{2-}$) or peroxymonosulfate (PMS, HSO_5) as oxidants. Sulfate radical has a half-life longer than ·OH (hydroxyl radicals); SO_4^- has a greater redox potential (2.5e3.1 V) than ·OH that has a standard redox potential of 1.8e2.7 V; and SR is more selective for the oxidation of organic contaminants over a wide pH range. Sulfate radicals react with organic molecules by removing hydrogen, adding to double bonds, and transferring electrons. Electron-donating groups will have a faster reaction rate than electron-withdrawing groups with the SR because SR is electrophilic. Some organics will react immediately with persulfate, creating SRs that propagate secondary reactions or organic radicals that decompose the desired pollutants [22]. Peroxydisulfate is not costly, and has high aqueous solubility and stability at room temperature, that is why it is often utilized as a source of SO_4^- [23, 24]. SBA-15 silica has been used as promising support for metal catalysts due to its outstanding hydrothermal and mechanical stabilities. Similarly, Fe- and Co-based catalysts supported on SBA-15 are widely used in catalytic degradation of non-biodegradable organic compounds [25]. Electro-activated persulfates could remove various POPs, such as pesticides, dyes, and pharmaceuticals, from simulated water at the laboratory scale. Wu and co-authors reported that the EC/Fe^{2+}/$S_2O_8^{2-}$ process removed 65.8% of acid orange 7 from wastewater in 60 min [26]. It has been shown that the combination of Electrochemical (EC) and heterogeneous activation of PDS using Fe–Co/SBA-15 as catalyst is a satisfactory technique for POPs removal—excellent in real water treatment (e.g., groundwater and wastewater) [27]. Electrochemical technique has recently been used in various investigations for persulfate activation due to its advantages in creating less sludge and so reducing both reactor volume and investment costs. Some refractory organic pollutants can be completely removed from contaminated sites under ideal conditions [28].

One of the homogenous phase methods of AOPs, the application of ionizing radiation, is considered as one of the most favorable and effective AOPs in the removal of POPs. Water and wastewater treatment with ionizing radiation is favorable because both e_{aq}^- and $^•OH$ are generated in the process of water radiolysis when diluted aqueous solutions (natural waters/most types of wastewaters of various origins) are irradiated. Depending on the chemical reactivity of the target species, water radiolysis products may participate in oxidative reactions with organic contaminants [29]. The most common process is water radiolysis, which produces compounds that react with dissolved species. Physical ($\approx$ 1 fs), physico-chemical ($10^{-15} – 10^{-12}$ s), which involves numerous processes, and chemical stage ($10^{-12} – 10^{-6}$ s) are the three primary stages that occur at distinct rates [30]. Ionization radiation such as electromagnetic and gamma ones, high energy electrons (electron beam, EB) and charged particles and neutrons, virtually only γ-rays, and EB irradiations are employed in water and wastewater treatment [31]. Untreated (standard blue) chemical effluent from dye manufactures treatment with EB irradiation was effective in significantly decreasing toxicity and color; $p < 0.0001$ was obtained for the sample treated by 2.5 kGy [32]. The hyphenation of established original AOPs with biological treatment for ordinary applications is one of the most recent advances [33]. A combination of electron-beam of radiation source 1 MeV EB accelerator, 400 kW and biological treatment was used for purification of dyeing complex wastewater under continuous flow conditions in South Korea by Han et al. [34].

Heterogeneous photocatalysis, one of the most AOPs for water purification, is an effective, cost-effective, and environmentally friendly method of eliminating organic pollutants. During the process, a semi-conductor irradiated with an appropriate wavelength of light produces active species, which oxidize organic compounds dissolved in water. The primary benefit of this approach is that it is inherently destructive; it does not require mass transfer; can be performed at ambient conditions (atmospheric oxygen is utilized as an oxidant); and can result in complete mineralization of organic carbon into CO_2 [35]. The pollutant, catalyst, and source of illumination must all be in close proximity or in contact for photocatalysis to work efficiently and effectively. Advanced oxidation technology's ability to remove minimal concentrations of POPs from water has been thoroughly demonstrated, and the technology is gradually being implemented in many parts of the world, including developing countries [36]. Semiconductor photocatalysts commonly used include TiO_2, CeO_2, ZnO, Fe_2O_3, CdS, and so on. Due to its strong photocatalytic capability, chemical and biological inertness, excellent photochemical stability, and inexpensive price, titanium dioxide (TiO_2) is a widely utilized photocatalyst in environmental degradation of organic molecules which led to complete mineralization into CO_2, H_2O, and harmless inorganic anions [37]. After 180 min of irradiation ($\lambda > 400$ nm), about 99% pentachlorophenol (PCP) was removed by Ag-deposited TiO_2 nanotubes (TNTs) under simulated solar light [38]. Several other photocatalysts have been utilized for removal of POPs in wastewater. Lopes da Silva et al. [39] investigated the degradation of perfluorooctanoic acid (PFOA) in water using Indium oxide and the results showed a good potential of nanosized In_2O_3 photocatalyst in degradation. Nanocrystalline ZnO particles doped with different concentrations of Fe impurity was able to degrade methylene blue (MB) dye in aqueous solution under UV/sunlight exposure [40]. Zn/TiO_2 catalyst synthesized from the hydrothermal method removed *ca.* 80% of paraquat from aqueous solution (using 4 g L^{-1} of catalyst) under UV and solar light irradiation [41].

Electrohydraulic discharge system is one of the most advanced oxidation methods for degrading hazardous organic contaminants in water and wastewater. Electrical plasma technology, as one of AOPs, has sparked a lot of interest in the removal of organic pollutants, owing to the absence of external chemicals, environmental compatibility, ability to kill microbes, non-generation of secondary pollution high removal efficiency, efficacy, and ease of operation at ambient temperature and pressure [42, 43]. Depending on the solution pH, conductivity, and discharge magnitude, an electrohydraulic discharge system can activate both the physical process and the chemical reaction mechanism, which subsequently generates free active species such as H_2O_2, OH radical, O, O_3, and $O_2\cdot^-$. This system could be combined with a number of AOPs including chemical, photolysis, ultrasonic irradiation, electrical, and supercritical water oxidation in water. Oxidative degradation of medicinal drug diclofenac (DCF) in water was investigated using a pulsed corona discharge generated above liquid. Efficient removal of DCF in water was achieved after 15 min of non-thermal plasma treatment as DCF in solution was totally removed [44]. Two different non-thermal plasma dielectric barrier discharge (DBD) reactors (planar and coaxial) at atmospheric pressure were assessed for the removal of organic micropollutants (Atrazine, Chlorfenvinfos, 2,4-Dibromophenol, and Lindane) from aqueous solutions (1–5 mg L^{-1}) at laboratory scale. The parent compounds disappeared as the plasma treatment time increased, and the degradations in both DBD devices followed first-order kinetics (k) in distilled water. The highest k value was recorded for 2,4-dibromophenol in the planar reactor, whereas the lowest k value was obtained for atrazine in the coaxial reactor [45]. The degradation of 13 distinct textile colors in an experimental DBD plasma batch reactor was investigated, during a 600-second treatment in the batch reactor, between 90 and 99% of most dyes were degraded. The investigation revealed that the DBD approach could be utilized to remediate a range of synthetic dye polluted water at low concentrations (up to 50 mg L^{-1}) with success [46].

Ozonation is another AOP that aims to degrade a wide range of organic pollutants in water by targeting their unsaturated hydrocarbon bonds [47]. Ozonation is the most promising method for pollutant degradation, according to laboratory and pilot-scale research, because it successfully eliminates a variety of substrates and by-products. Degradation, by ozonation, of 20 mg L^{-1} of sulfadimethoxine (SDM) at pH 7.0 in different water matrices was investigated. Water treatment *via* ozonation was proven to be effective as 100% removal of SDM was achieved within 10 min [48]. Dar et al. [49] used two different ozone generators (sources: water and air) to test the degradation of pentabromophenol (PBP) in an aqueous media. The water-source ozone generator achieved complete degradation of 50 μmol L^{-1} PBP after 5 min, and the air-source ozone generator achieved complete degradation of 10 μmol L^{-1} PBP after 45 min. The authors found out that ozonation is an effective and suitable procedure for PBP degradation in real water systems. The effectiveness of ozone treatment to eliminate the 16 priority Polycyclic aromatic hydrocarbons (PAHs) in waste activated sludge (WAS) was investigated by optimizing ozonation performance by varying key operating variables, including ozone gas flow rate, inlet concentration and dosage and then explore the pH dependent behavior of ozone-oxidation. The PAHs removal efficiency increased with ozone dosage and was strongly pH dependent. Even at ozone dosage of 40 mg $O_3 \cdot g^{-1}$, the PAHs removal efficiency at pH 9.0 (44.5%) was significantly higher than the one observed at pH 5.0 and 200 mg $O_3 \cdot g^{-1}$ (41.7%). The research indicates the need of WAS disintegration during ozonation to make PAHs more accessible to O_3 molecules and ·OH to initiate oxidation reactions and recommended to adopt a sequential batch operation for ozonation to mitigate the negative effect of soluble organic compounds generated by sludge solubilization so as to practically use ozonation for elimination of PAHs in WAS [50]. Nanotechnology can also be incorporated into ozonation for more efficient removal. The elimination of five organochlorine pesticides {hexachlorobutadiene (HCHBD), pentachlorobenzene (PCHB), hexachlorobenzene (HCHB), lindane (LIN), and heptachlor (HCH)} using integrated O_3/nZVI procedures in water solution was explored. Except for LIN and HCHB, the ozonation method showed high removal efficiency of >90% after 60 min. The O_3/UV procedure yielded somewhat higher removal rate and efficiency. Within 5 min of starting the nZVI procedure, high removal efficiency for LIN, PCHB, and HCH were measured. The findings imply that the O_3/nZVI mechanisms have great potential for increasing organochlorine pollutants breakdown and elimination [51]. Ozone-based processes, i.e., single ozonation, O_3/UVA, O_3/UVA/Fe^{3+} and O_3/UVA/magnetite have been shown to be suitable technologies to deal with POPs [52].

Fenton oxidation is an AOP that is cost-effective and efficient method of removing POPs from water because of the low toxicity of the reagents, absence of mass transfer limitation due to its homogeneous catalytic nature (i.e., Fe^{2+} and H_2O_2), and the simplicity of the technology. The standard Fenton reaction, on the other hand, has a number of drawbacks, such as a narrow pH range, the formation of Fe-containing sludge, and a low hydrogen peroxide usage rate [53]. The Fenton system involves combining ferrous ions with hydrogen peroxide to produce hydroxyl radicals, which have a strong oxidizing ability and can breakdown organic pollutants. Ferric ions are generated during the reaction, which can be reacted to yield ferrous ions. A Fenton-like reaction is a reaction that occurs when hydrogen peroxide reacts with ferric ions. The main disadvantage of this method is the high cost of the reactants, H_2O_2 and $Fe^{2+.}$ As a result, numerous methods have been developed to utilize Fe^{3+} salts rather than Fe^{2+} salts, resulting in the photo-Fenton and electro-Fenton approaches [54]. Amorphous FeOOH quantum dots (QDs) were coupled with polymeric photocatalysts *g*-C_3N_4 which was developed as a visible light driving photo-Fenton catalyst. Highly dispersed FeOOH QDs anchored on *g*-C_3N_4 showed enhanced visible light driving photo-Fenton degradation of methylene orange (MO) and phenol—an indication that the FeOOH QDs coupled with g-C_3N_4 is a promising visible light driving photo-Fenton catalyst for organic pollutants treatment. S-doped NiFe-based particles were prepared by a solvothermal method and used

POPs	AOPs	Remarks	References
Pharmaceuticals (clofibric acid, carbamazepine, iomeprol)	Heterogeneous photocatalysis	HPLC/DAD/FLD were used to measure the degradation of the persistent contaminants. When it comes to clofibric acid breakdown, TiO2 photocatalyst P25 outperformed the Hombikat UV100. A better choice than P25 for photocatalytic degradation of iomeprol was the Hombikat UV100. The photocatalysis of clofibric acid was generally slowed down by the presence of natural organic matter and carbamazepine. As a result of this research, it is clear that photocatalysis is an effective way to eliminate POPs, even when they are present at low concentrations or in a complex matrix.	[57]
Para-chlorophenol	Heterogeneous catalytic ozonation	In the investigation, para-chlorophenol (4-CP) was subjected to ozonation and hydrogen peroxide was produced simultaneously. When molecular ozone attacks the double bonds of 4-CP's aromatic ring, hydrogen peroxide is formed, which partially deprotonated to HO2 in the neutral pH range. The HO2 accelerated the breakdown of ozone to become ·OH radicals. The ·OH probe chemical succinic acid was created. As the probe chemical vanished, it confirmed the existence of the OH radical dot radical, and its dispersion across time. Ozone and the produced ·OH enabled the breakdown of 4-CP into small molecular weight organic compounds without scavengers inactivating the radicals.	[58]
20 organic contaminants (tetrahydrofuran, benzene, pyridine, ethylbenzene, 3-methylpyridyne, toluene, furfural, dibutyl sulfide, o-xylene, o-cresol, m-cresol, phenol, nitrophenol, tert-butyl disulfide, 2,4-dimethylpyridine, 2,4,6-trimethylpyridine, 4-methylbenzaldehyde, 2,4-ethylphenol, naphthalene and p-nitrotoluene)	Hydrodynamic cavitation	A number of investigations were conducted using hydrodynamic cavitation to remediate a model effluent with 20 organic compounds representing various industrial contaminants. Three mineral acids were used to acidify the wastewater. For acidification, sulfuric acid was the only chemical that resulted in effective treatment without the production of secondary pollutants that could be observed in the field. When cavitation was used for 6 h, sulfuric acid gave the best treatment, with a total organic carbon (TOC) removal rate of 90% in the majority of cases. Nitric acid was less efficient (> 60% for most of the compounds). Hydrochloric acid had the lowest performance, with most compounds degrading at <50%.	[59]
Perfluorooctanesulfonate (PFOS)	Ionizing radiation	In aqueous solution using an electron beam, the work examined the effect of oxidants on radiolysis degradation of PFOS and the mechanism for decomposition of PFOS. In spite of the fact that the absorbed dose increased, PFOS breakdown efficiency dropped. It was shown that PFOS breakdown efficiency was roughly the same when using an electron beam plus 5.0 mM persulfate at a dose 100 kGy, as compared to using only electron beam at the same dosage.	[60]
Phenol	Fenton reaction	Thermal shrinkage polymerization was used to produce Fe-g-C3N4 *in situ*. Under visual irradiation, the photocatalysis-Fenton system with H2O2 showed excellent phenol removal rate. Fenton-photocatalysis eliminated 20 ppm of phenol in 50 min. The investigators investigated the most effective conditions to remove phenol: catalyst Fe-g-C3N4 with a 5% doping, an 8 mM H2O2 concentration, a pH of 3, and catalyst dose of 1.5 g L^{-1}.	[61]

POPs	AOPs	Remarks	References
Phenol	Ozonation	Studies were conducted on the effects of operating parameters such as pH and ozone dosage. The breakdown of contaminants increased with increasing pH (3–11). As the ozone dosage rate rose (from 5.5 to 36.17 mg L^{-1} min^{-1}), so did the efficiency of pollutant removal. As a result of the ozonation process, the wastewater's ultraviolet absorbance (UV 254 nm) reduced, indicating the breakdown of complex organic molecules into low molecular weight organic compounds. Water's pH dipped from 11 to roughly 8.5 as a result of the formation of intermediate acidic species in the process.	[62]
Pentachlorophenol (PCP)	Photo-Fenton	It has been shown that PCP (1 mg L^{-1}) undergoes photo-Fenton degradation in the presence of simulated and natural sun irradiation. In situations when PCP photolysis is important, the soluble bio-based substances (SBO) screen has a negative impact on PCP degradation. In contrast, the photo-Fenton process was significantly improved when SBO was introduced without PCP photolysis. As a result of SBO's capacity to complex iron and prevent its precipitation as oxides or hydroxides, photo-Fenton technique can be used at pH values that are close to neutral. A larger concentration of Fe(II) (4 mg L^{-1}) showed a favorable effect on PCP degradation, while at 1 mg L^{-1}, PCP degradation rates were equivalent in the presence and absence of SBO.	[63]

Table 3.
POPs removal using various AOPs.

for degradation of methylene blue (MB) from aqueous solutions with visible light in photo Fenton reaction. Results showed that $NiFe_2S_4$ has a great performance of MB degradation in the photo-Fenton oxidation process; 100 mL of 30 mg L^{-1} MB could be completely degraded (99.8%) within 6 min under optimal reaction conditions [55]. Xiang and co-authors employed yolk-shell $ZnFe_2O_4$ as photo-Fenton catalyst to investigate antibiotics degradation. The yolk-shell $ZnFe_2O_4$ not only exhibited excellent photocatalytic activity for the removal of popular pollutants, but also for the co-existing pollutants consisted of tetracycline (TC) and Ciprofloxacin (CIP). They reported a novel technique for preparing high efficiency of photo-Fenton catalysts for decontamination of refractory pollutants in aqueous solutions [56].

Various studies conducted on the usage of AOPs for elimination of POPs from water are summarized in **Table 3**.

4. Conclusion

Persistent organic pollutants get into the environment through municipal populations, industrial and agricultural activities. A number of chemicals have been identified as POPs, and these chemicals include, but not limited to, Aldrin, Endrin, Heptachlor, Mirex, Chlordane, Toxaphene, Dieldrin, Hexachlorobenzene, Furans (polychlorinated dibenzofurans), Polychlorinated biphenyls, Dioxins (polychlorinated dibenzo-p-dioxins) and DDT. Advanced oxidative processes (AOPs) are aqueous phase oxidation systems that have high ability to eliminate POPs from water systems. Although several advanced oxidation processes (AOPs) have recently achieved success in the treatment of POPs in wastewater, the successes of POPs treatment using various advanced technologies are not without downsides, such as low degradation efficiency, toxic intermediate generation, massive sludge production, high energy expenditure and high operational cost. Combination of AOPs is recommended for effective elimination of POPs from water.

Conflict of interest

The authors declare no conflict of interest.

Author details

Ifeoluwa Oluwafunmilayo Daramola[1] and Matthew Ayorinde Adebayo[2*]

1 Department of Chemistry, University of Fort Hare, Alice, South Africa

2 Department of Chemistry, The Federal University of Technology, Akure, Nigeria

*Address all correspondence to: adebayoma@futa.edu.ng

References

[1] Alharbi OML, Arsh A, Khattab RA, Ali I. Health and environmental effects of persistent organic pollutants. Journal of Molecular Liquids. 2018;**263**:442-453. DOI: 10.1016/j.molliq.2018.05.029

[2] Vallack HW et al. Controlling persistent organic pollutants – what next ? 1998;**6**:143-175. DOI: 10.1016/S1382-6689(98)00036-2

[3] Pariatamby A, Kee YL. Persistent organic pollutants management and remediation. Procedia Environmental Sciences. 2016;**31**:842-848. DOI: 10.1016/j.proenv.2016.02.093

[4] Jeon J, Murugesan K, Nam I, Chang Y. Coupling microbial catabolic actions with abiotic redox processes: A new recipe for persistent organic pollutant (POP) removal. Biotechnology Advances. 2013;**31**(2):246-256. DOI: 10.1016/j.biotechadv.2012.11.002

[5] Abdel-shafy HI, Mansour MSM. A review on polycyclic aromatic hydrocarbons: Source, environmental impact , effect on human health and remediation. Egyptian Journal of Petroleum. 2016;**25**(1):107-123. DOI: 10.1016/j.ejpe.2015.03.011

[6] Jones KC, De Voogt P. Persistent organic pollutants (POPs): state of the science. 1999;**100**:209-221. DOI: 10.1016/S0269-7491(99)00098-6

[7] Adithya S, Sai R, Krishnan A, Malolan R, Govarthanan M. A critical review on the formation, fate and degradation of the persistent organic pollutant hexachlorocyclohexane in water systems and waste streams. Chemosphere. 2021;**271**:129866. DOI: 10.1016/j.chemosphere.2021.129866

[8] Ma J, Hung H, Macdonald RW. The influence of global climate change on the environmental fate of persistent organic pollutants: A review with emphasis on the Northern Hemisphere and the Arctic as a receptor. Global and Planetary Change. 2016;**146**: 89-108. DOI: 10.1016/j.gloplacha.2016.09.011

[9] Adithya S et al. A critical review on the formation, fate and degradation of the persistent organic pollutant hexachlorocyclohexane in water systems and waste streams. Chemosphere. 2021;**271**:129866. DOI: 10.1016/j.chemosphere.2021.129866

[10] Ritter L, Solomon KR, Forget J, Stemeroff M, Group TC. A review of selected Heptachlor-Hexachlorobenzene-Mïrex-toxaphene polychlorinated biphenyls dioxins and furans. Prepared for the International Programme on Chemical Safety within the framework of the Inter-Organization Programme for the Sound Management of Chemicals. 1995. pp. 1-149. Accessed on August 5, 2021 from https://www.epa.gov/sites/production/files/2015-01/documents/rmpp_6thed_final_lowresopt.pdf

[11] Wu J. Soil-air partition coefficients of persistent organic pollutants decline from climate warming: A case study in Yantai County, Shandong Province, China. Water, Air, and Soil Pollution. 2020;**231**(7):371-395. DOI: 10.1007/s11270-020-04718-4

[12] Wang XP, Sun DC, Yao TD. Climate change and global cycling of persistent organic pollutants: A critical review. Science China Earth Sciences. 2016; **59**(10):1899-1911. DOI: 10.1007/s11430-016-5073-0

[13] Gascon M, Morales E, Sunyer J, Vrijheid M. Effects of persistent organic pollutants on the developing respiratory and immune systems: A systematic review. Environment International. 2013;**52**:51-65. DOI: 10.1016/j.envint.2012.11.005

[14] Nobile M et al. Presence of emerging contaminants in baby food. Food Additives & Contaminants: Part A: Chemistry, Analysis, Control, Exposure & Risk Assessment. 2020;**37**(1):131-142. DOI: 10.1080/19440049.2019.1682686

[15] Scheringer M. Long-range transport of organic chemicals in the environment. Environmental Toxicology and Chemistry. 2009;**28**(4):677-690. DOI: 10.1897/08-324R.1

[16] Bouwman H. South Africa and the Stockholm convention on persistent organic pollutants. South African Journal of Science. 2004;**100**(7-8): 323-328

[17] Environmental Protection Agency Ghana (EPA). Accra, Ghana: Ghana National Implementation Plan of the Stockholm Convention on Persistent Organic Pollutants. 2007. pp. 1-267

[18] Badmus KO, Tijani JO, Massima E, Petrik L. Treatment of persistent organic pollutants in wastewater using hydrodynamic cavitation in synergy with advanced oxidation process. Environmental Science and Pollution Research. 2018;**25**(8):7299-7314. DOI: 10.1007/s11356-017-1171-z

[19] Kasprzyk-Hordern B, Dinsdale RM, Guwy AJ. The removal of pharmaceuticals, personal care products, endocrine disruptors and illicit drugs during wastewater treatment and its impact on the quality of receiving waters. Water Research. 2009;**43**(2):363-380. DOI: 10.1016/j.watres.2008.10.047

[20] Bethi B et al. Investigation of TiO2 photocatalyst performance for decolorization in the presence of hydrodynamic cavitation as hybrid AOP. Ultrasonics Sonochemistry. 2016;**28**:150-160. DOI: 10.1016/j.ultsonch.2015.07.008

[21] Arzate S, Pfister S, Oberschelp C, Sánchez-Pérez JA. Environmental impacts of an advanced oxidation process as tertiary treatment in a wastewater treatment plant. Science of the Total Environment. 2019;**694**: 133572. DOI: 10.1016/j.scitotenv.2019.07.378

[22] Tsitonaki A, Petri B, Crimi M, Mosbk H, Siegrist RL, Bjerg PL. In situ chemical oxidation of contaminated soil and groundwater using persulfate: A review. Critical Reviews in Environmental Science and Technology. 2010;**40**(1):55-91. DOI: 10.1080/10643380802039303

[23] Wang X, Wang L, Li J, Qiu J, Cai C, Zhang H. Degradation of acid orange 7 by persulfate activated with zero valent iron in the presence of ultrasonic irradiation. Separation and Purification Technology. 2014;**122**:41-46. DOI: 10.1016/j.seppur.2013.10.037

[24] Guan YH et al. Efficient degradation of atrazine by magnetic porous copper ferrite catalyzed peroxymonosulfate oxidation via the formation of hydroxyl and sulfate radicals. Water Research. 2013;**47**(14):5431-5438. DOI: 10.1016/j.watres.2013.06.023

[25] Zhong X et al. Mesoporous silica iron-doped as stable and efficient heterogeneous catalyst for the degradation of C.I. Acid Orange 7 using sono-photo-Fenton process. Separation and Purification Technology. 2011;**80**(1):163-171. DOI: 10.1016/j.seppur.2011.04.024

[26] Wu J, Zhang H, Qiu J. Degradation of acid orange 7 in aqueous solution by a novel electro/Fe 2+/peroxydisulfate process. Journal of Hazardous Materials. 2012;**215-216**:138-145. DOI: 10.1016/j.jhazmat.2012.02.047

[27] Cai C, Zhang H, Zhong X, Hou L. Electrochemical enhanced heterogeneous activation of

peroxydisulfate by Fe-Co/SBA-15 catalyst for the degradation of Orange II in water. Water Research. 2014;**66**:473-485. DOI: 10.1016/j.watres.2014.08.039

[28] Matzek LW, Carter KE. Activated persulfate for organic chemical degradation: A review. Chemosphere. 2016;**151**:178-188. DOI: 10.1016/j.chemosphere.2016.02.055

[29] Trojanowicz M, Bobrowski K, Szreder T, Bojanowska-Czajka A. Chapter 9 - Gamma-ray, X-ray and electron beam based processes. In: Ameta SC, Ameta R, editors. *Advanced Oxidation Processes for Waste Water Treatment*. Cambridge, United States: Academic Press; 2018. pp. 257-331

[30] Le Caër S. Water radiolysis: Influence of oxide surfaces on H2 production under ionizing radiation. Water (Switzerland). 2011;**3**(1):235-253. DOI: 10.3390/w3010235

[31] Trojanowicz M. Removal of persistent organic pollutants (POPs) from waters and wastewaters by the use of ionizing radiation. Science of the Total Environment. 2020;**718**:134425. DOI: 10.1016/j.scitotenv.2019.134425

[32] Borrely SI, Morais AV, Rosa JM, Badaró-Pedroso C, da Conceição Pereira M, Higa MC. Decoloration and detoxification of effluents by ionizing radiation. Radiation Physics and Chemistry. 2016;**124**:198-202. DOI: 10.1016/j.radphyschem.2015.11.001

[33] Charlesby A, Davison WHT, Lloyd DG. Intensity and concentration dependence of some radiation-induced reactions in anthracene solutions. The Journal of Physical Chemistry. 1959;**63**(6):970-975. DOI: 10.1021/j150576a046

[34] Han B et al. Combined electron-beam and biological treatment of dyeing complex wastewater. Pilot plant experiments. Radiation Physics and Chemistry. 2002;**64**(1):53-59. DOI: 10.1016/S0969-806X(01)00452-2

[35] Konstantinou IK, Albanis TA. TiO2-assisted photocatalytic degradation of azo dyes in aqueous solution: Kinetic and mechanistic investigations: A review. Applied Catalysis B: Environmental. 2004;**49**(1):1-14. DOI: 10.1016/j.apcatb.2003.11.010

[36] Ibhadon AO, Fitzpatrick P. Heterogeneous photocatalysis: Recent advances and applications. Catalysts. 2013;**3**(1):189-218. DOI: 10.3390/catal3010189

[37] Li F et al. Hydrothermal synthesis of graphene grafted titania/titanate nanosheets for photocatalytic degradation of 4-chlorophenol: Solar-light-driven photocatalytic activity and computational chemistry analysis. Chemical Engineering Journal. 2017, 2018;**331**:685-694. DOI: 10.1016/j.cej.2017.09.036

[38] Yu L, Yang X, Ye Y, Peng X, Wang D. Silver nanoparticles decorated anatase TiO2 nanotubes for removal of pentachlorophenol from water. Journal of Colloid and Interface Science. 2015;**453**:100-106. DOI: 10.1016/j.jcis.2015.04.057

[39] Lopes da Silva F, Laitinen T, Pirilä M, Keiski RL, Ojala S. Photocatalytic degradation of perfluorooctanoic acid (PFOA) from wastewaters by TiO2, In2O3 and Ga2O3 catalysts. Topics in Catalysis. 2017;**60**(17-18):1345-1358. DOI: 10.1007/s11244-017-0819-8

[40] Kumar K, Chitkara M, Sandhu IS, Mehta D, Kumar S. Photocatalytic, optical and magnetic properties of Fe-doped ZnO nanoparticles prepared by chemical route. Journal of Alloys and Compounds. 2014;**588**:681-689. DOI: 10.1016/j.jallcom.2013.11.127

[41] Sakee U, Wanchanthuek R. Catalytic activity of bimetallic Zn/{TiO}2 catalyst

for degradation of herbicide paraquat: Synthesis and characterization. Materials Research Express. 2017; **4**(11):115504. DOI: 10.1088/2053-1591/aa8e59

[42] Locke BR, Sato M, Sunka P, Hoffmann MR, Chang J-S. Electrohydraulic discharge and nonthermal plasma for water treatment. Industrial and Engineering Chemistry Research. 2006;**45**(3):882-905. DOI: 10.1021/ie050981u

[43] Jiang B et al. Review on electrical discharge plasma technology for wastewater remediation. Chemical Engineering Journal. 2014;**236**:348-368. DOI: 10.1016/j.cej.2013.09.090

[44] Dobrin D, Bradu C, Magureanu M, Mandache NB, Parvulescu VI. Degradation of diclofenac in water using a pulsed corona discharge. Chemical Engineering Journal. 2013;**234**:389-396. DOI: 10.1016/j.cej.2013.08.114

[45] Hijosa-Valsero M, Molina R, Schikora H, Müller M, Bayona JM. Removal of priority pollutants from water by means of dielectric barrier discharge atmospheric plasma. Journal of Hazardous Materials. 2013;**262**:664-673. DOI: 10.1016/j.jhazmat.2013.09.022

[46] Tichonovas M et al. Degradation of various textile dyes as wastewater pollutants under dielectric barrier discharge plasma treatment. Chemical Engineering Journal. 2013;**229**:9-19. DOI: 10.1016/j.cej.2013.05.095

[47] Masten SJ, Davies SHR. The use of ozonation to degrade organic contaminants in wastewaters. Environmental Science & Technology. 1994;**28**(4):180A-185A. DOI: 10.1021/es00053a718

[48] Shad A, Li C, Zuo J, Liu J, Dar AA, Wang Z. Understanding the ozonated degradation of sulfadimethoxine, exploration of reaction site, and classification of degradation products. Chemosphere. 2018;**212**:228-236. DOI: 10.1016/j.chemosphere.2018.08.050

[49] Dar AA et al. Ozonation of pentabromophenol in aqueous basic medium: Kinetics, pathways, mechanism, dimerization and toxicity assessment. Chemosphere. 2019;**220**: 546-555. DOI: 10.1016/j.chemosphere.2018.12.154

[50] Li W, Zhu N, Shen Y, Yuan H. Towards efficient elimination of polycyclic aromatic hydrocarbons (PAHs) from waste activated sludge by ozonation. Environmental Research. 2020, 2021;**195**:110783. DOI: 10.1016/j.envres.2021.110783

[51] Derco J, Dudáš J, Valičková M, Šimovičová K, Kecskés J. Removal of micropollutants by ozone based processes. Chemical Engineering and Processing Process Intensification. 2015;**94**:78-84. DOI: 10.1016/j.cep.2015.03.014

[52] Espejo A, Beltrán FJ, Rivas FJ, García-Araya JF, Gimeno O. Iron-based catalysts for photocatalytic ozonation of some emerging pollutants of wastewater. Journal of Environmental Science and Health - Part A Toxic/Hazardous Substances and Environmental Engineering. 2015;**50**(6):553-562. DOI: 10.1080/10934529.2015.994939

[53] Bel Hadjltaief H, Sdiri A, Ltaief W, Da Costa P, Gálvez ME, Ben Zina M. Efficient removal of cadmium and 2-chlorophenol in aqueous systems by natural clay: Adsorption and photo-Fenton degradation processes. Comptes Rendus Chimie. 2018;**21**(3-4):253-262. DOI: 10.1016/j.crci.2017.01.009

[54] Wang S. A Comparative study of Fenton and Fenton-like reaction kinetics in decolourisation of wastewater. Dyes and Pigments. 2008;**76**(3):714-720. DOI: 10.1016/j.dyepig.2007.01.012

[55] Guo X, Meng Q, Wang D, Zha F, Tang X, Tian H. S-doped NiFe-based catalyst for fast degradation of methylene blue by heterogeneous photo-Fenton reaction. Environmental Science and Pollution Research. 2021;**28**:36112-36121. DOI: 10.1007/s11356-021-13212-1

[56] Xiang Y, Huang Y, Xiao B, Wu X, Zhang G. Magnetic yolk-shell structure of ZnFe2O4 nanoparticles for enhanced visible light photo-Fenton degradation towards antibiotics and mechanism study. Applied Surface Science. 2020;**513**:145820. DOI: 10.1016/j.apsusc.2020.145820

[57] Doll TE, Frimmel FH. Removal of selected persistent organic pollutants by heterogeneous photocatalysis in water. Catalysis Today. 2005;**101**(3-4) SPEC. ISS.:195-202. DOI: 10.1016/j.cattod.2005.03.005

[58] Pi Y, Zhang L, Wang J. The formation and influence of hydrogen peroxide during ozonation of para-chlorophenol. Journal of Hazardous Materials. 2007;**141**(3):707-712. DOI: 10.1016/j.jhazmat.2006.07.032

[59] Gągol M, Cako E, Fedorov K, Soltani RDC, Przyjazny A, Boczkaj G. Hydrodynamic cavitation based advanced oxidation processes: Studies on specific effects of inorganic acids on the degradation effectiveness of organic pollutants. Journal of Molecular Liquids. 2020;**307**:113002. DOI: 10.1016/j.molliq.2020.113002

[60] Kim TH, Lee SH, Kim HY, Doudrick K, Yu S, Kim SD. Decomposition of perfluorooctane sulfonate (PFOS) using a hybrid process with electron beam and chemical oxidants. Chemical Engineering Journal. 2018, 2019;**361**:1363-1370. DOI: 10.1016/j.cej.2018.10.195

[61] Hu J, Zhang P, An W, Liu L, Liang Y, Cui W. In-situ Fe-doped g-C3N4 heterogeneous catalyst via photocatalysis-Fenton reaction with enriched photocatalytic performance for removal of complex wastewater. Applied Catalysis B: Environmental. 2018, 2019;**245**:130-142. DOI: 10.1016/j.apcatb.2018.12.029

[62] Mecha AC, Onyango MS, Ochieng A, Momba MNB. Impact of ozonation in removing organic micro-pollutants in primary and secondary municipal wastewater: Effect of process parameters. Water Science and Technology. 2016;**74**(3):756-765. DOI: 10.2166/wst.2016.276

[63] Vergura EP, Garcia-Ballestreros S, Vercher RF, Santos-Juanes L, Prevot AB, Arques A. Photo-fenton degradation of pentachlorophenol l: Competition between additives and photolysis. Nanomaterials. 2019;**9**(8):1-8. DOI: 10.3390/nano9081157